Applications of Optical Fibers for Sensing

Edited by Christian Cuadrado-Laborde

Published in London, United Kingdom

IntechOpen

Supporting open minds since 2005

Applications of Optical Fibers for Sensing
http://dx.doi.org/10.5772/intechopen.78283
Edited by Christian Cuadrado-Laborde

Contributors
Jonas Osório, Cristiano Cordeiro, Sanjay Kher, Manoj Kumar Saxena, Domenico Lo Presti, Giuseppe Gallo, Danilo Luigi Bonanno, Daniele Bongiovanni, Fabio Longhitano, Santo Reito, Dinesh Babu Duraibabu, Gabriel Leen, Fintan McGuinness, Elfed Lewis, Gerard Dooly, Daniel Toal, Rebecca Yen-Ni Wong, Dora Juan Juan Hu, Perry Ping Shum, Morten Ibsen, Martina Delgado-Pinar, Xavier Roselló-Mechó, Antonio Diez, Jose Luis Cruz, Miguel V. Andrés, Emmanuel Rivera, Maria Fátima Domingues, Cátia Tavares, Tiago Leite, Catia Leitao, Nélia Alberto, Carlos Marques, Ayman Radwan, Eduardo Rocon, Paulo Fernando Da Costa Antunes, Paulo André, Christian A Cuadrado-Laborde

First published in London, United Kingdom, 2019 by IntechOpen
IntechOpen is the global imprint of INTECHOPEN LIMITED, registered in England and Wales, registration number: 11086078, The Shard, 25th floor, 32 London Bridge Street
London, SE19SG – United Kingdom
Printed in Croatia

British Library Cataloguing-in-Publication Data
A catalogue record for this book is available from the British Library

Additional hard copies can be obtained from orders@intechopen.com

Applications of Optical Fibers for Sensing
Edited by Christian Cuadrado-Laborde
p. cm.
Print ISBN 978-1-78985-351-3
Online ISBN 978-1-78985-352-0

For EU product safety concerns:
IN TECH d.o.o., Prolaz Marije Krucifikse Kozulić 3, 51000 Rijeka, Croatia, info@intechopen.com or visit our website at intechopen.com.

Meet the editor

Christian Cuadrado-Laborde received his PhD degree in Physics from the University of La Plata (UNLP, Argentina, 2005), and his Electrical and Electronic Engineering degree from the University of San Luis (UNSL, Argentina, 1998). He is an independent researcher of CONICET (2005), the most important federal research agency in Argentina. He leads the Laboratory of Optics and Photonics in the Institute of Physics Rosario (IFIR, Rosario, Argentina). He is currently also Full Professor of the Engineering Faculty of the Pontifical Catholic University of Argentina (Rosario, Argentina).
He was awarded by the Spanish Ministry of Science with a research fellow position to join the Optical Fiber Laboratory group headed by M.V. Andrés of the University of Valencia (Spain) during 2008–2009. Since then, he has continued collaborating actively with this research group. He has published several research papers in top-tier scientific journals such as *Progress in Optics*, *Optics Letters*, *Optics Express*, *Laser & Photonics Reviews*, *Optics & Photonics News*, etc. His current research interest includes fiber optics applications, photonic signal processing, and mode-locked all-fiber lasers.

Contents

Preface

I have been involved in the research field of fiber optics for more than 15 years by working in all fiber lasers, all-optical signal processing, etc. Therefore, and considering that the field of optical fibers is within my research area of expertise, I accepted the very kind invitation of the IntechOpen editorial house to edit a book in this special field. Today, it is a great honor and pleasure for me to introduce this new book: *Applications of Optical Fibers for Sensing*. This is my third collaboration as book editor with IntechOpen after the successful experiences of *Applications of Digital Signal Processing* (2011) and *Current Trends in Short- and Long-Period Fiber Gratings* (2012). I truly hope that this new book will also be successfully received by the worldwide scientific community.

In this book the reader will find a collection of chapters written by different experts around the world, covering the broad field of sensing of different parameters through optical fibers. The main goal of this book is to give an updated overview of the current research trends in this rapidly evolving field.

First, in Chapter 2, "Minimalist Approach for the Design of Microstructured Optical Fiber Sensors," J.H. Osório and C.M.B. Cordeiro from Brazil report on the utilization of capillary-like fibers of different kinds as platforms for the realization of sensing measurements. These authors show that external stimuli can alter fiber wall thickness and/or induce birefringence variations, which in turn allows its use as a pressure sensor, temperature sensor, or refractive index sensor. Next, in Chapter 3, "Fiber Bragg Gratings as e-Health Enablers: An Overview for Gait Analysis Applications," M.F. Domingues et al. from Portugal and Spain report on the use of fiber Bragg gratings for gait analysis monitoring, i.e. the monitoring of plantar pressure, angle, and torsion, with the purpose of future integration in rehabilitation exoskeletons and for prosthetic control.

In Chapter 4, "Distributed, Advanced Fiber Optic Sensors," S. Kher and M.K. Saxena from India report on several optical fiber sensing techniques such as Raman scattering and fiber Bragg gratings, to be applied especially in the context of nuclear plant monitoring. In Chapter 5, "Real-Time Particle Radiography by Means of Scintillating Fibers Tracker and Residual Range Detectors," D. Lo Presti et al. from Italy report in detail on the construction and characterization of a charged particle imaging system by using scintillating optical fibers.

In Chapter 6, "Review of Liquid-Filled Optical Fibre-Based Temperature Sensing," F. McGuinness et al. from Ireland review on the current state of the art in liquid-filled optical fibre temperature sensing, with the focus on high-resolution temperature sensing. In Chapter 7, "Optical Fibre Long-Period Grating Sensors Operating at and around the Phase Matching Turning Point," R.Y.-N. Wong et al. from Singapore and the United Kingdom report on the behavior and potential uses for fiber optic sensing of a very special type of very sensitive long-period fiber grating: those working at (and around) the phase matching turning point.

Finally, in Chapter 8, "Whispering Gallery Modes for Accurate Characterization of Optical Fibers' Parameters," M. Delgado-Pinar et al. from Spain review the use of a special technique to measure temperature profiles along conventional and special fibers (such as photosensitive or doped fibers), elasto-optic coefficients, and ultraviolet-induced absorption loss coefficients of different photosensitive fibers. This technique makes use of whispering gallery modes, which are resonant in optical wavelength; being their spectral position a function of the radius and the refractive index of the microresonator material. Due to the high-quality factor of these resonances, they allow different parameters with high sensitivities and very low detection limits to be measured.

I would like to thank all the authors for their contributions, since their support has been crucial for the fulfillment of this project. In addition, I would like to thank the IntechOpen editorial staff for the confidence placed in me to edit this book, and especially to IntechOpen's Author Service Manager Sara Debeuc for her support during the whole process. On behalf of myself and the authors, we hope readers enjoy this book and benefit from its content, which provides a thorough understanding of several fields related to this exciting discipline.

Christian Cuadrado-Laborde
Instituto de Física Rosario (CONICET-UNR),
Pontificia Universidad Católica Argentina,
Facultad de Química e Ingeniería,
Rosario, Argentina

Chapter 1

Introductory Chapter: Application of Optical Fiber for Sensing

Christian Cuadrado-Laborde

1. Introduction

The history of the use of optical fiber for sensing applications began with two different, but interrelated, discoveries: laser light and optical fibers. The first laser was built in 1960 by T. H. Maiman at Hughes Research Laboratories, based on the theoretical work by C. H. Townes and A. L. Schawlow. A laser provides a source of an intense coherent light, highly collimated, and quasi-monochromatic; its potential for data transfer was immediately envisaged. Naturally, first experiments involved the transmission of the laser beam through the air. However, a communication channel cannot be practically sustained propagating freely through the air, owing to atmospheric attenuation and weather influence. Researchers also conducted experiments by transmitting the laser beam through glass fibers, which soon became the preferred medium for transmission of light. First, optical fibers were not practical to sustain a communication channel mainly due to the presence of impurities in the fiber material, resulting in very high transmission losses (>1000 dB/km), until Corning presented at the beginning of the 1970s optical fibers with (in comparison) very lower transmission losses, with only a few dB/km. Today, typical transmission losses are below 0.2 dB/km. This represents an extraordinary improvement as compared with electrical signal transmission through coaxial cables, not to mention the wider bandwidth available, which is several orders of magnitudes higher.

These developments paved the way to a plethora of different works on fiber optic sensing. But, what is an optical fiber sensor? **Figure 1** shows a block diagram of a typical optical fiber sensor. It is composed of a light source (which not only can be a laser, but also a broad band light source like a light emitting diode, etc.), the optical fiber itself transmits the light from the light source to the sensing area

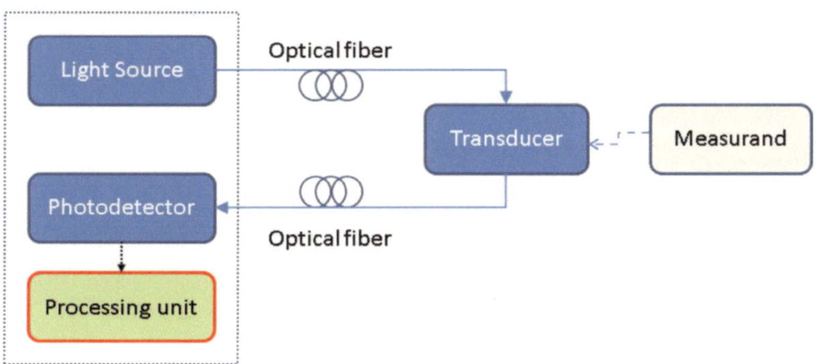

Figure 1.
General block diagram for an optical fiber sensor.

(forward optical fiber), the sensing element (which can be the same optical fiber, or another specially designed optical fiber, or not fiber at all, see below the differences between intrinsic and extrinsic optical fiber detectors), another optical fiber, which carry the light with the information of the measurand of interest (in some case, backward or return optical fiber; in this case, forward and backward fibers are the same fiber; it typically used some additional optical fiber element like an optical circulator), a photo-detector, which converts light in an electrical signal, and an end-processing device or processing unit (which converts this electric signal, eventually digitized in the measurand of interest; in a research laboratory environment, it would be an optical-spectrum analyzer, oscilloscope, etc.; and in a final user environment, the processing unit is simplified in order to provide more easy access to the measurand information, and of course, in order to reduce costs).

Compared with other types of sensors, fiber optic sensors exhibit a number of advantages, among them are the following:

- Possibility of use in high-voltage environments, since an optical fiber is made fundamentally of silica, which is an electrically insulating material.

- Possibility of use in explosive environments, since electricity is not present, there is no risk at all of electrical sparks.

- Optical fibers are chemically passive (the object to be measured can hardly be contaminated).

- Immunity to corrosion.

- Immunity to electromagnetic interference, no matter how much intense they were. Because an optical fiber is electrically nonconductive, it becomes impossible to act as an antenna.

- Very wide operating temperature range.

- Easy integration into a wide variety of structures, including composite materials, with little interference due to their small size and cylindrical geometry.

- Lightweight, which is fundamental in aero spatial applications.

- Robust and resistant to harsh environments.

- Material cost is relatively low, which in turn prevents theft (as compared with copper wires).

- Security of information passed down the cable.

- Highly sensitive sensing capabilities.

- Multiplexing capability to form sensing networks, multiple sensors in a single fiber can be simultaneously interrogated.

- Remote sensing capability.

- Multifunctional sensing capabilities such as strain, pressure, corrosion, temperature, and acoustic signals.

- No need for power amplifiers in a broad length range, since the attenuation losses can be very low, even over long distances.

Optical fiber sensor can be divided into intrinsic or extrinsic. In the first, intrinsic, the measurand of interest interacts itself with the optical fiber; that is, the optical fiber structure is modified under the influence of the measurand and the fiber itself plays an active role in the sensing function. As a consequence, there is some kind of modulation of light inside the fiber (this modulation can affect the intensity, wavelength, phase, or polarization), which carries the information of the performed measurement. Thus, by detecting these parameters and their changes, the external perturbations can be sensed. This is the reason why they are also called all-fiber sensors. A typical example of this kind of sensors is a fiber Bragg grating working as a strain sensor; under different strains, the Bragg reflection shifts to a different wavelength. On the other hand, in an extrinsic sensor, the measurand does not act on the fiber itself, it acts outside the fiber. In this case, the optical fiber merely acts as a transmission medium, and of course light collection. Some fiber optic Fabry-Perot interferometers are good examples of this kind.

The aforementioned division is not unique, and optical fiber sensors can be divided according to other characteristics. One especially useful is according to the point of measurement. Therefore, we can distinguish single-point sensing, quasi-distributed sensing, and continuous distributed sensing. In the former, there is one single point of measurement; for example, a LPG measuring the concentration of some organic compound at a specific place. On the contrary, in quasi-distributed sensing, there are integer numbers of point sensors along the optical fiber; for example, when several fiber Bragg gratings are employed as sensing elements distant some length between them (in this case, some kind of multiplexing is necessary). Finally, in continuous distributed sensing, there is

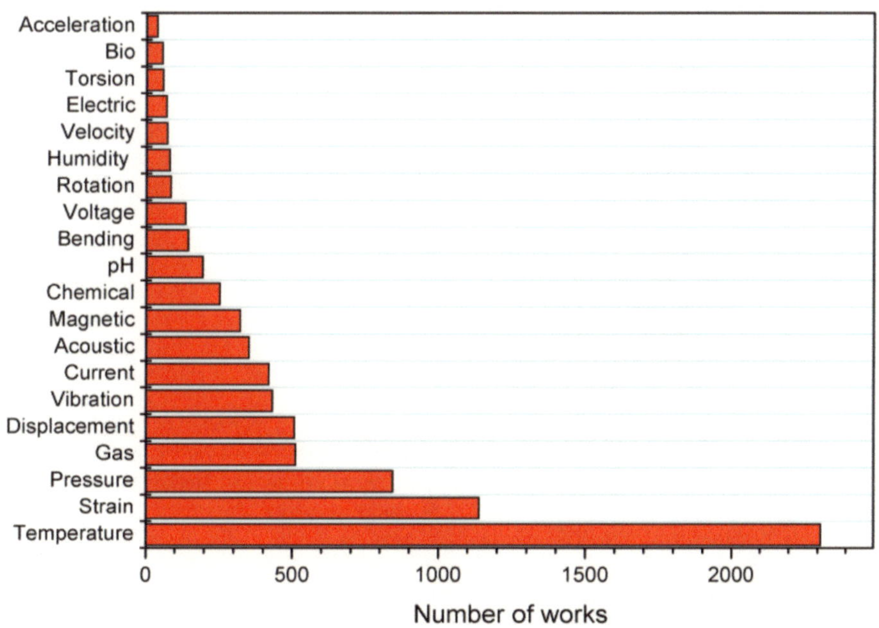

Figure 2.
Published works along all years distributed according to the nature of the measurands. Patents are not included. In those cases, where the work deals with more than one measurand; for example, temperature and strain, the work counts for both.

a continuous, real-time measurement along the entire length of an optical fiber cable; not only there are theoretically infinite numbers of measurement points, but also the optical fiber itself is the sensing element (as an example in Raman distributed optical sensing).

Optical fiber sensors have attracted the attention of several research groups around the world along all these years. **Figure 2** shows the result of a rapid search of those published works with the simultaneous presence in their titles of three words: fiber + sensor + measurand of interest (i.e., temperature, strain, pressure, gas concentration, etc.). Temperature and strain are the most frequent physical quantities measured through optical fiber sensors, which is not surprising at all, since first efforts were focused in this direction.

The optical fiber sensors in turn, can be constructively performed in several ways, either by using: short (Bragg) or long period gratings, interferometers (Fabry-Perot, Mach-Zehnder, etc.), by detecting different scattered light signals (Raman, Brillouin, and Rayleigh), photonic crystal fibers, surface plasmon resonance (SPR) etc. **Figure 3** shows the number of those published works with the simultaneous presence in their titles of three words: fiber + sensor + operation principle (i.e., Bragg grating, photonic crystal fiber, Fabry-Perot, etc.). It should not be surprising that the most used fiber optic sensor is the fiber Bragg grating, which is motivated firstly by its versatility.

To resume, the optical fiber sensing technology is today a successful and mature technology after a half-century of development [1–5], with a year-to-year increasing market participation. It is expected that in the near future arose new emerging applications offering new opportunities for research and exploitation.

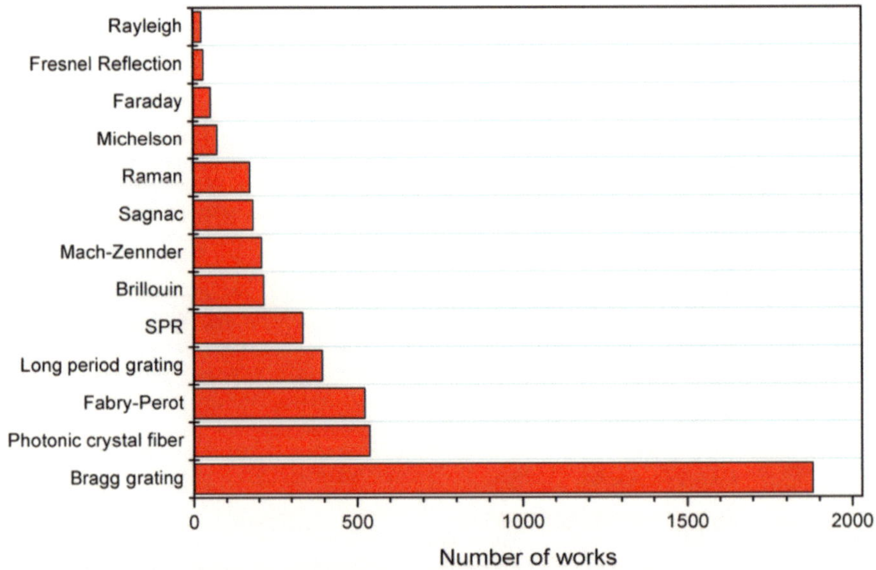

Figure 3.
Published works along all years distributed according to the operation principle. Patents are not included. In those cases, where the work deals with more than one operation principle (e.g., fiber Bragg grating and long-period grating), the work counts for both.

DOI: http://dx.doi.org/10.5772/intechopen.83623

Author details

Christian Cuadrado-Laborde[1,2]*

1 Instituto de Física Rosario (CONICET-UNR), Rosario, Argentina

2 Facultad de Química e Ingeniería, Pontificia Universidad Católica Argentina, Rosario, Argentina

*Address all correspondence to: christian.cuadrado@uv.es

References

[1] Lee B. Review of the present status of optical fiber sensors. Optical Fiber Technology. 2003;**9**(2):57-79. DOI: 10.1016/S1068-5200(02)00527-8

[2] Lopez-Higuera JM, editor. Handbook of Optical Fibre Sensing Technology. New York: John Wiley & Sons Inc.; 2002. 828 pp. ISBN 0-47182-053-9

[3] Di Sante R. Fibre optic sensors for structural health monitoring of aircraft composite structures: Recent advances and applications. Sensors. 2015;**15**(8):18666-18713. DOI: 10.3390/s150818666

[4] Culshaw B. Optical fibre sensors: A current perspective. The Open Optics Journal. 2013;7(Suppl-1, M2):21-31. DOI: 10.2174/1874328501307010021

[5] Culshaw B, Kersey A. Fiber-optic sensing: A historical perspective. Journal of Lightwave Technology. 2008;**26**(9):1064-1078. DOI: 10.1109/JLT.0082.921915

Chapter 2

Minimalist Approach for the Design of Microstructured Optical Fiber Sensors

Jonas H. Osório and Cristiano M. B. Cordeiro

Abstract

We report on recent investigations regarding ultra-simplified designs for microstructured optical fiber sensors. This minimalist approach relies on the utilization of capillary-like fibers—namely embedded-core fibers, surface-core fibers, and capillary fibers—as platforms for the realization of sensing measurements. In these fibers, guidance of light is accomplished in an embedded or surface germanium-doped core or in the hollow part of capillaries. External stimuli can alter fiber wall thickness and/or induce birefringence variations, allowing, for the embedded-core and capillary fibers, to operate as pressure or temperature sensors. For the surface-core fiber design, the interaction between the guided mode and external medium allows the realization of refractive index sensing either by using fiber Bragg gratings or surface plasmon resonance phenomenon. Also, we report the realization of directional curvature sensing with surface-core fibers making use of the off-center core position. The attained sensitivities are comparable to the ones obtained with much more sophisticated structures. The results demonstrate that these novel geometries enable a new route toward the simplification of optical fiber sensors.

Keywords: fiber optics, fiber optics sensors, fiber Bragg gratings, surface plasmon resonance, microstructured optical fibers

1. Introduction

The increasing need for the development of optical sensors motivates intense research in this area. Particularly, great efforts have been observed in the field of optical fiber sensors since they can provide numerous advantages such as high sensitivity and improved resolution. Moreover, fiber optics are immune to electromagnetic interference and suitable to be used in harsh environments.

In this context, microstructured optical fibers have much contributed to the development of sensors due to their huge design freedom. Thus, numerous sensing configurations have been reported in the literature to be able to probe variations of a great diversity of parameters such as temperature, strain, hydrostatic pressure, curvature, and refractive index. Regarding pressure sensors, for example, successful approaches are to use photonic-crystal fibers (PCFs) with triangular-shaped microstructures [1] or side-hole PCFs [2]. In these configurations, the hydrostatic pressure application generates asymmetric stresses distributions within the fiber structure and changes its birefringence via the photoelastic effect. Additionally, metal-filled side-hole PCFs have been demonstrated to act as high-sensitivity temperature sensors [3, 4]. In this approach, the metal expansion inside the fiber structure induces an asymmetric stress distribution within the same (due to the different thermal expansion coefficients of silica and of the metal). It entails changes in the fiber birefringence which can be optically probed and related to the temperature variation through a suitable calibration.

Microstructured optical fibers can also be used as refractive index sensors. Among those ones, plasmonic sensors acquire great importance due to the high sensitivities they can achieve. In these platforms, selected regions of the fiber structure are coated with a nanometric-thick metallic layer to provide coupling between the optical mode and a plasmonic mode. Possible approaches are to coat the inner holes of microstructured fibers [5] or to open up a channel in the fiber structure to expose the fiber core for metallic nanospheres immobilization [6].

However, the microstructured optical fibers usually employed in the sensing schemes described above are sophisticated, which demand great technical efforts for their fabrication. Here, alternatively, we present sensors which are endowed with ultra-simplified microstructures based on capillary-like fibers (embedded-core fibers [7, 8], surface-core fibers [9, 10], and capillary fibers [11]). As it will be shown in the following, even though these configurations are very simple, the attained sensitivities are high when compared to other fiber sensors based on more complex structures. Thus, we can identify the use of capillary-like fibers as a new avenue for obtaining highly sensitive fiber sensors with simplified fabrication process.

2. Capillary-like fiber designs for sensing applications

Figure 1 shows the fiber designs we present here. In **Figure 1a**, a diagram of the embedded-core fiber is shown [7, 8]. This structure consists of a silica capillary endowed with a germanium-doped core, which is placed inside the wall of the capillary. The embedded-core fiber can be employed for pressure or temperature sensing measurements. Specifically, when the embedded-core fiber acts as a temperature sensor, the capillary hollow part must be filled with metal (**Figure 1b**). The principle of operation of these sensors is centered on the capillary wall displacements that occur when the hollow embedded-core fiber is pressurized or when the metal-filled embedded-core fiber experiences temperature variations. These wall displacements within the capillary wall entail asymmetric stress distributions in the

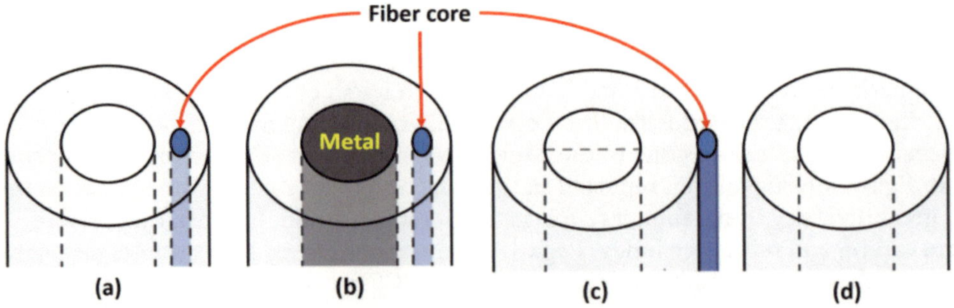

Figure 1.
(a) Embedded-core fiber, (b) metal-filled embedded-core fiber, (c) surface-core fiber and (d) capillary fiber.

fiber structures, which generate, by virtue of the photoelastic effect, birefringence variations in the fiber core.

Figure 1c shows a diagram of the so-called surface-core fiber [9]. In this structure, the fiber core is placed on the fiber external surface. As here the core directly interfaces the external medium, the evanescent field of the guided optical mode permeates the external environment. Surface-core fibers are, then, a suitable platform for refractive index sensing. A possible approach is to imprint fiber Bragg gratings (FBGs) in the fiber core and measure the sensor spectral response as the external refractive index is altered [9]. A second approach is to perform plasmonic sensing by metal-coating the surface-core fiber with a nanometric metallic layer [10]. Additionally, the off-center position of the fiber core permits the surface-core fibers to act as directional curvature sensors. In this case, the curvature-induced strain levels within the core can be probed by a FBG [9].

Figure 1d presents a forth structure, which is simply a capillary fiber [12]. Here, we study the guidance of light in the hollow part of the capillary and investigate how the optical response of the fiber is changed when it experiences temperature variations. It is worth saying that, in this investigation, we employed polymethyl methacrylate (PMMA) capillaries. This choice was due to the higher thermal expansion and thermo-optic coefficients of PMMA when compared to silica.

In the next sections, we specifically describe the principle of operation of each configuration. Moreover, we present theoretical and experimental results and compare them with the ones available in the literature.

3. Embedded-core capillary fibers for pressure sensing

The application of pressure to capillary fibers generates displacements on their walls. This, in turn, induces an asymmetric stress distribution within the capillary structure which, due to the stress-optic effect, entails birefringence variations in it. As described in [7], an analytical model can be used to account for the material birefringence variations (ΔB_{mat}) in pressurized capillaries. To do that, we can employ Eq. (1), where C_1 and C_2 are the elasto-optic coefficients ($C_1 = -0.69 \times 10^{-12}$ and $C_2 = -4.19 \times 10^{-12}$ Pa^{-1} for silica), and σ_x and σ_y are the pressure-induced stresses on the horizontal and vertical directions, respectively [13, 14].

$$\Delta B_{mat} = (C_2 - C_1)\left(\sigma_x - \sigma_y\right) \tag{1}$$

The stresses σ_x and σ_y can be obtained from Lamé solution inside thick-walled tubes subjected to pressure [15]. The resulting expression for the material birefringence at a position x on the horizontal axis is shown in Eq. (2), where r_{in} and r_{out} are the inner and outer radius of the capillary, and $p_{gauge} = p_{out} - p_{in}$ (p_{in} and p_{out} are the inner and outer pressure levels) [7].

$$\Delta B_{mat} = 2(C_2 - C_1)p_{gauge}\left[1 - \left(\frac{r_{in}}{r_{out}}\right)^2\right]^{-1}\frac{r_{in}^2}{x^2} \tag{2}$$

By observing Eq. (2), we see that, when maintaining r_{in} constant, $|\Delta B_{mat}|$ will be greater for higher r_{in}/r_{out} ratios. It means that the analytical model predicts that the change in the birefringence is increased for thin-walled capillaries. Moreover, we see that, for fixed r_{in} and r_{out} values, the change in the birefringence is higher for positions (x) which are closer to the inner wall of the capillary [7].

Although the analytical model for the material birefringence can provide important information on the most important geometrical parameters that affects the sensitivity of the sensor, it is necessary to account for the modal birefringence dependence on the applied pressure for a broader understanding of the sensor characteristics. To do this, a numerical simulation of the embedded-core fiber structure was carried on. **Figure 2** presents the results for dB_{modal}/dP (derivative of the modal birefringence as a function of the pressure) as a function of the core position within the capillary. In the simulations, r_{in} = 40 μm and r_{out} = 67.5 μm. The core dimensions were considered to be 5.7 and 11.4 μm.

The results presented in **Figure 2** show, as could be expected from the analytical model, that dB_{modal}/dP values are higher for core positions which are closer to the inner wall of the capillary. However, we note that, very interestingly, the trend that is expected from the analytical model is verified only when the whole core is within the capillary wall. When the core has part of its area outside of the capillary wall, a strong decrease in dB_{modal}/dP is observed (core region is represented as dark blue ellipses in **Figure 2**). This allows observing that, for maximizing dB_{modal}/dP in embedded-core fibers, it is crucial that the whole core is inside the capillary wall.

In order to obtain an experimental demonstration of the proposed design acting as a pressure sensor, we performed the fabrication of the embedded-core fiber. The fabrication process is simple and with few steps. Initially, a germanium-doped silica rod is merged to a silica tube. In sequence, the resulting preform is inserted in another silica tube, which acts as a jacket. The preform is then drawn in a fiber tower facility [7]. **Figure 3a** shows the cross-section of the embedded-core fiber.

Figure 3b exposes a diagram of the experimental setup used for pressure measurements. Light from a broadband source (BLS) is launched in the fiber and detected with an optical spectrum analyzer (OSA). Polarizers (P_1 and P_2) are used to excite the orthogonal modes of the fiber and to recombine them after traveling along the fiber. A pressure chamber is used to subject the fiber to different pressure levels.

By using the configuration of **Figure 3b**, an interference spectrum is measured in the OSA. Since the embedded-core fiber is sensitive to pressure variations, the spectral position of the interferometric fringes is shifted when the external pressure level is altered. A sensitivity coefficient, C_S, is defined to account for the spectral shift of the fringes as a function of the pressure variation, $\frac{d\lambda_{IF}}{dP}$. The C_S value can also be written as a function of the wavelength, λ, the fiber group birefringence, G, and

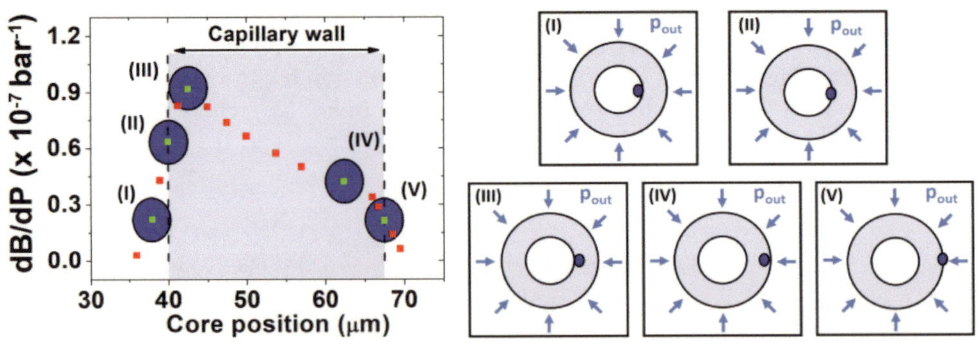

Figure 2.
Modal birefringence derivative as a function of pressure for different core positions inside the capillary wall. Light blue region represents the capillary wall, and dark blue ellipses represent the core region. Insets illustrate the core position within the capillary.

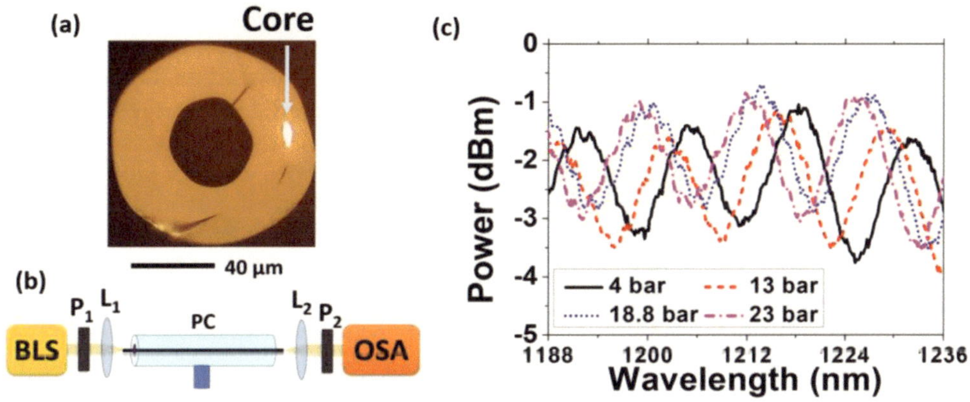

Figure 3.
(a) Embedded-core fiber. (b) Experimental setup for pressure sensing measurements. BLS: broadband light source. P_1 and P_2: polarizers. L_1 and L_2: objective lenses. PC: pressure chamber. (c) Spectral response of the embedded-core fiber for different pressure levels.

of the modal phase birefringence derivative with respect to the pressure, $\frac{\partial B_{modal}}{\partial P}$ — Eq. (3) [2].

$$C_S \equiv \frac{d\lambda_{IF}}{dP} = \frac{\lambda}{G}\frac{\partial B_{modal}}{\partial P} \tag{3}$$

Figure 3c presents the measured optical response of the embedded-core fiber for different pressurization conditions. It is seen that when the pressure level increases, the fringes blueshift. After performing an appropriate correction on the fiber pressurized and nonpressurized lengths [2, 16], we can estimate a sensitivity value of (1.04 ± 0.01) nm/bar [7]. This value is high when compared to other results measured in polarimetric measurements reported in the literature. For example, in [17], a sensitivity of 0.342 nm/bar was measured for a commercial photonic-crystal fiber. Additionally, in [18, 19], the sensitivities of 0.30 and 0.52 nm/bar were reported for specially designed microstructured fibers.

Moreover, the $\frac{\partial B_{modal}}{\partial P}$ value for the embedded-core fiber can be estimated from Eq. (3). The resulting value is $(2.33 \pm 0.02) \times 10^{-7}$ bar^{-1}, which is in the same magnitude order of the ones found for much more sophisticated microstructured fibers [1]. Therefore, we see that the embedded-core fiber structure allows achieving high sensitivity (C_s) and $\frac{\partial B_{modal}}{\partial P}$ even with a nonoptimized fiber. It also allows recognizing embedded-core fiber as a promising platform for the realization of pressure sensing using optical fibers and a novel route for the design of microstructured optical fiber pressure sensors.

4. Embedded-core capillary fibers for temperature sensing

Embedded-core fibers can also act as highly sensitive temperature sensors if a metal is inserted into the hollow part of the capillary (**Figure 1b**). In analogy to the embedded-core fiber pressure sensor, the principle of operation is based on the induction of stresses inside the capillary and on the consequent variation of the fiber birefringence [8].

In this configuration, the metal expansion inside the capillary causes its volume elements to displace and to compress the silica capillary structure. This, in turn, entails an asymmetric stress distribution within the capillary wall and induces birefringence variations. An analytical model can be used to predict the most relevant parameters that contribute to the sensor response. This analytical model provides Eq. (4), which accounts for the variation in the material birefringence of the capillary, ΔB_{mat}, for a given temperature variation, ΔT (r_{in} and r_{out} are the capillary inner and outer radii, and C_1 and C_2 are the silica elasto-optic coefficients). The parameter δ in Eq. (4) is given by Eq. (5), where υ is the Poisson ratio, E is the Young modulus, and α is the thermal expansion coefficient. In Eq. (5), index 1 refers to the filling metal and index 2 refers to silica [8]. By observing Eq. (4) and Eq. (5), it is possible to realize that $|\Delta B_{mat}|$ will be greater for positions closer to the inner radius and when the filling metal has a larger thermal expansion coefficient.

$$\Delta B_{mat} = -2\delta\Delta T(C_2 - C_1)\frac{r_{out}^2}{x^2} \tag{4}$$

$$\delta = \frac{(1+\nu_2)\alpha_2 - (1+\nu_1)\alpha_1}{\frac{(1+\nu_1)}{E_1}\left(\nu_1 - \frac{1}{2}\right) + \frac{(1+\nu_2)}{E_2}\left(\nu_2 - \frac{1}{2} - \frac{r_{out}^2}{r_{in}^2}\right)} \tag{5}$$

Thus, indium was chosen to be the filling metal due to its high thermal expansion coefficient (32.1×10^{-6}°C^{-1}) and reasonably low melting point (156°C) [3]. The low melting point is an important property since it simplifies the metal filling process. To insert the metal inside the embedded-core fiber, the metal is molten and pressure is applied to push it into the hollow region.

Figure 4a shows the indium-filled embedded-core fiber. To measure its temperature sensitivity, the same experimental setup as represented in **Figure 3b** was used. The unique difference is that, for the temperature sensing measurements, the pressure chamber was substituted by a water reservoir placed on a hot plate in order to adequately alter the fiber temperature.

Figure 4b shows the spectra measured for different temperature conditions. We see that there is a spectral shift toward longer wavelengths as the temperature is increased. After performing a suitable correction on the heated and unheated fiber lengths, the sensitivity was calculated to be (8.40 ± 0.06) nm/°C [8]. This sensitivity value compares well to the highest temperature sensitivity values reported in the literature such as 9.0[4], 6.6 [20], and 16.49 nm/°C [21], which were measured for photonic-crystal fibers filled with indium, ethanol, and index matching liquid, respectively. This once again demonstrates that embedded-core fibers are a very promising platform for the realization of high-sensitivity optical sensing.

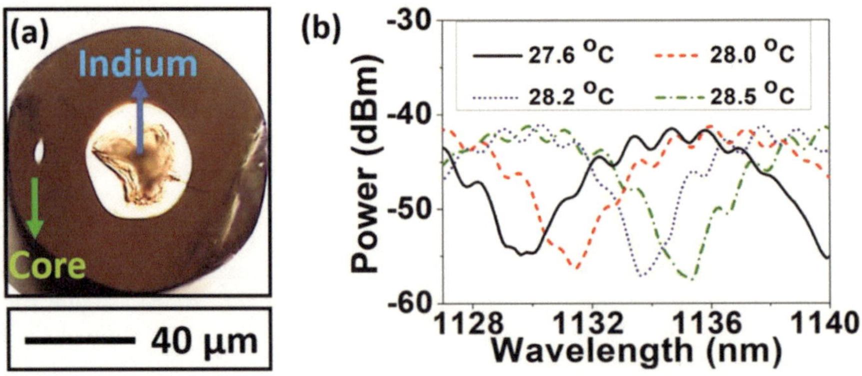

Figure 4.
(a) Indium-filled embedded-core fiber. (b) Spectral response of the indium-filled embedded-core fiber for different temperature conditions.

5. Surface-core fibers for refractive index sensing

A third sensing opportunity under the approach of simplified optical fiber sensors is the employment of surface-core fibers [9]. In this kind of fibers, the core is placed on the outer surface of the same (**Figure 1c**). The proximity of the core to the external environment makes surface-core fibers suitable to be used as refractive index sensors, and the off-center position of the core allows these fibers to operate as directional curvature sensors. **Figure 5a** presents the cross-section of the surface-core fiber.

In order to explore the refractive index sensitivity of surface core fibers, fiber Bragg gratings (FBGs) can be inscribed in the fiber core [9]. FBGs consist of a refractive index modulation in the core of the fiber able to couple the propagating core mode to a contra-propagating one. The coupling between the modes is achieved at a specific wavelength, λ_B, which can be accounted by Eq. (6), where n_{eff} is the effective refractive index of the core mode and Λ is the pitch of the grating. Experimentally, the response of FBGs is seen as a reflection peak at λ_B.

$$\lambda_B = 2n_{eff}\Lambda \tag{6}$$

As the optical mode guided in the core directly interfaces the external medium, the effective refractive index of the core mode will be dependent on the external refractive index variations. Thus, if the refractive index of the external environment is altered, a shift in the spectral position of the Bragg peak is expected.

Figure 5b presents the measured wavelength shift of the Bragg peak as a function of the external refractive index. As can be seen in **Figure 5b**, the results for the surface-core fiber showed low sensitivity. To improve the sensor response, tapers from the surface-core fiber were prepared prior to grating inscription. By doing this, it is possible to enhance the interaction between the guided mode evanescent

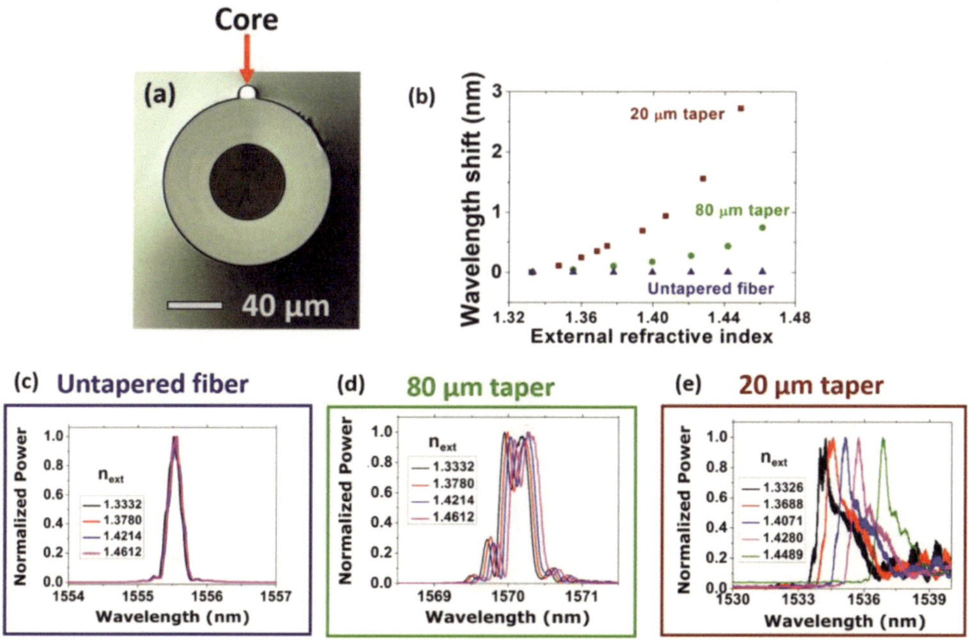

Figure 5.
(a) Surface-core fiber. (b) Wavelength shift of the Bragg peak as a function of the external refractive index. Reflected Bragg peaks in the (c) untapered surface-core fiber and in the (d) 80 μm and (e) 20 μm tapers.

field and the external medium and improve the sensitivity. The results measured for tapers with 80 and 20 μm diameter are also shown in **Figure 5b**. The Bragg peaks for the untapered surface-core and for the 80 and 20 μm tapers are shown in **Figure** 5c–e—the gratings were imprinted via phase-mask technique by employing phase masks with pitches 1075.34 nm (for the 80 μm taper) and 1071.2 nm (for the untapered fiber and for the 20 μm taper). Around 1.42 RIU (refractive index unit), the measured sensitivities were 8 and 40 nm/RIU for the 80 and 20 μm tapers, respectively. These sensitivity values compare well with other results for FBG-based refractive index sensors. In [22], it is reported a sensitivity of 15 nm/RIU for a Bragg grating inscribed in a 6–μm-thick taper (measured around 1.326 and 1.378 RIU) and, in [23], a sensitivity of 30 nm/RIU was measured for a 8.5-μm taper (in the same refractive index range). It is worth observing that the results for surface-core fibers were attained for thicker tapers, which implies in sensor robustness improvement.

Another possibility for exploring the sensitivity of surface-core fibers to variations in the refractive index of the external medium is to functionalize the core with a metallic nanolayer and make the fiber to act as a plasmonic sensor. In this approach, the core mode can be resonantly coupled to a plasmonic mode when phase-matching occurs between them (surface plasmon resonance—SPR). The coupling between these modes is seen as a spectral dip at the wavelength where the phase matching condition is achieved. As the fiber core directly interfaces the external medium, the spectral position of the plasmonic resonance will be dependent on the refractive index of the external environment. The sensitivity of the configuration is, therefore, accounted from the spectral shift of the plasmonic resonance as a function of the external refractive index variation.

Figure 6a presents a simulation on the transmittance of the surface-core fiber plasmonic sensor for different refractive index values. In the simulations, the fiber core is assumed to have a 10 μm diameter and to be coated with a gold layer of 50 nm thick. It is seen that the plasmonic resonance is shifted toward longer wavelengths as the external refractive index increases. The sensitivity accounted from the simulations is 1290 nm/RIU. In addition, in **Figure 6b**, the core mode intensity profiles are shown when it is not phase-matched (off-resonance) and when it is phase-matched (in resonance) with the plasmonic mode. We can observe that, when the modes are in resonance, they hybridize. The intensity profiles in **Figure 6b** were calculated for an external refractive index of 1.39. The phase-matched intensity profile was accounted at 600 nm and the one for the off-resonance condition was accounted at 720 nm.

To experimentally test the proposed sensor, the fiber core was coated with a gold layer of 50 nm thick, and sensing measurements were performed at the Aalto

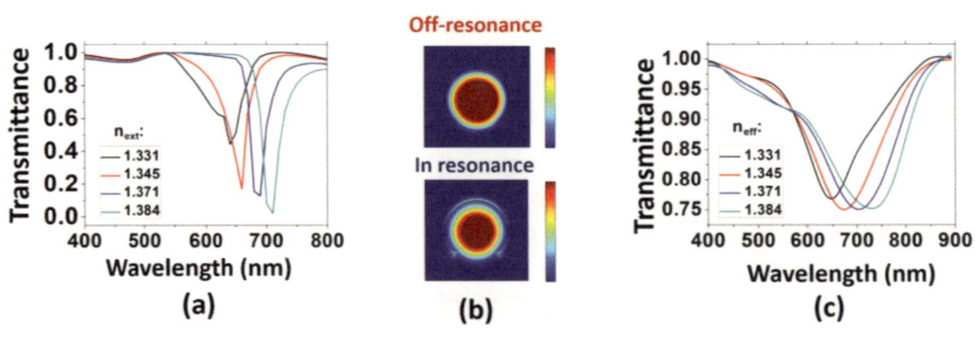

Figure 6.
(a) Simulated transmittance for different external refractive indexes (n_{ext}). (b) Intensity profiles of the core mode when off-resonance and when in resonance with the plasmonic mode. (c) Experimental transmittance for different external refractive indices.

University (Finland) [10] by immersing the sensor into glycerol-water solutions at different concentrations. **Figure 6c** shows the transmittance spectra as a function of the external refractive index. By following the spectral position of the spectral dip, we could measure a sensitivity of 1380 nm/RIU. This sensitivity value is comparable to the ones reported for other plasmonic sensors, which employ fragile fiber tapers and more sophisticated microstructured optical fibers [24, 25]. Thus, we can visualize that the plasmonic sensor based on a gold-coated surface-core fiber is a powerful platform for the realization of highly sensitive refractive index sensing. Moreover, the setup presents the advantages of increased robustness when compared to fragile fiber tapers and simpler preparation than the sensors, which demand metal coating of the inner holes of microstructured fibers.

6. Surface-core fibers for directional curvature sensing

Besides refractive index sensing using Bragg gratings or surface plasmon resonance, surface-core fibers also offer the possibility of the realization of directional curvature sensing. This is possible because the off-center position of the fiber core allows it to experience compression or expansion depending on the curvature direction. In this context, a FBG can be inscribed in the core of the surface-core fiber for probing the bend-induced strain levels inside the core and determine the direction of the curvature.

Compression and expansion of the core introduces strain levels in it. The induced strain in a bent fiber, ε, is proportional to the curvature, C (≡1/curvature radius). If the core is at a distance y from the fiber neutral axis, the induced strain level can be calculated from Eq. (7) [26]. Since the existence of a strain level in a fiber entails variations in its refractive index (by virtue of the strain optic effect) and length, the spectral response of a FBG in this fiber is expected to shift when it is subjected to strain increments. Eq. (8) describes the dependence of the Bragg peak shift as a function of the curvature. In Eq. (8), P_ε is the photoelastic coefficient of silica (P_ε = 0.22) [27].

$$\varepsilon = y\ C \tag{7}$$

$$\Delta\lambda_B = (1 - P_\varepsilon)\lambda_B\varepsilon = [(1 - P_\varepsilon)\lambda_B y]C \tag{8}$$

To experimentally test the response of the proposed sensor, a FBG was imprinted in the surface-core fiber (by using a phase mask with 1071.2 nm pitch) and the fiber was subjected to curvature increments. The results of the curvature sensing measurements are exposed in **Figure 7**. It is seen that when the fiber experiences expansion, the Bragg peak spectral position redshifts (positive spectral shift). Otherwise, when the fiber is compressed by the bending, the Bragg peak blueshifts (negative spectral shift). The measured sensitivities were (188 ± 5) and (202 ± 5) pm/m^{-1} for the expansion and compression conditions, respectively.

The achieved sensitivity values are high when compared to other FBG-based sensors whose performance is reported in the literature (sensitivities from 50 to 100 pm/m^{-1}) [28–30]. Additionally, the sensitivity for the surface-core fiber sensor is greater than the one obtained for FBGs sensors in eccentric core polymer optical fibers [31]. It is worth saying, however, that greater sensitivity values can be attained in other configurations. For example, we find in the literature that fibers with two or three cores can provide sensitivities of hundreds of nanometers per inverse meter [32, 33]. Nevertheless, the spectral features whose spectral shifts are considered in [32, 33] are much broader than the Bragg peak in the surface-core fiber we measured. This has an important impact on the sensor resolution. For the

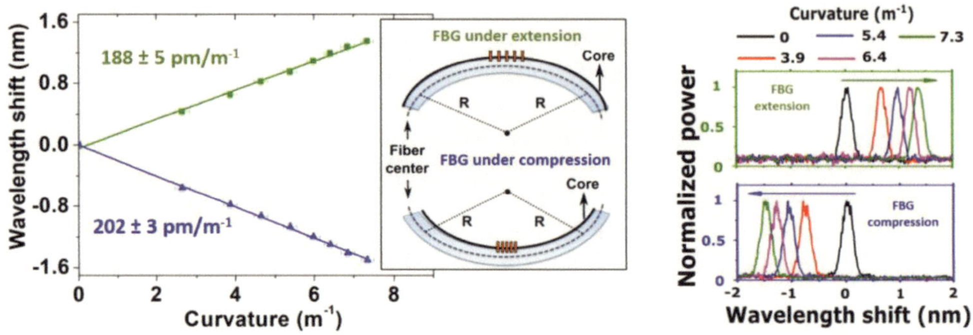

Figure 7.
Wavelength shift as a function of the curvature, representation of the curvature direction and spectral response of the directional curvature sensor based on a FBG inscribed in a surface-core fiber.

sensor reported in [32], for example, one can estimate a resolution limit of 0.01 m^{-1}, which is similar to the one we can find in our results (0.02 m^{-1}).

7. Polymer capillary fibers for temperature sensing

A simpler structure, which can be employed in sensing measurements , is capillary fibers [12]. In our approach, we investigated the sensitivity of polymer capillary fibers to temperature variations by studying the spectral characteristics of the light that is transmitted through its hollow part.

The typical transmission spectra through the hollow part of capillaries have wavelengths of high loss, λ_{min}. These wavelengths encounter high leakage during propagation because they are resonant with the capillary wall. At these wavelengths, which are given by Eq. (9), transmission minima are observed. In Eq. (9), n_1 is the refractive index of the hollow core, n_2 is the refractive index of the capillary material, d is the thickness of the capillary wall, and m is the order of the minimum. Eq. (9) tells that if the thickness and the refractive index of the capillary are altered, the minima spectral locations are expected to shift. As temperature variations are able to change both these parameters, capillary fibers can act as temperature sensors.

$$\lambda_{min} = \frac{2n_1 d}{m}\sqrt{\left(\frac{n_2}{n_1}\right)^2 - 1} \tag{9}$$

Taking this into account, polymethyl methacrylate (PMMA) capillaries were chosen to be used in this investigation. This decision was due to the high thermal expansion coefficient of PMMA (2.2×10^{-4}°C^{-1}) and its high thermo-optic coefficient (-1.3×10^{-4}°C^{-1}).

An analytical model can be used to investigate the influences of the thermal expansion and of the thermo-optic effect in the sensor [11, 34]. Eq. (10) presents an expression for calculating the output power (P_{out}) as a function of the wavelength (λ). In Eq. (10), P_{in} is the input power, L is the fiber length, d is the capillary wall thickness, n_1 and n_2 are refractive indexes of the hollow core and of the capillary material, and the parameter Γ is given by Eq. (11). In Eq. (10) and Eq. (11), θ_1 is the angle of incidence of the light rays on the capillary wall for a specific leaky mode—given by $\theta_1 = \sin^{-1}\left(\frac{n_{eff}}{n_1}\right)$. In the expression for θ_1, the effective refractive index of the leaky mode guided in the core, n_{eff}, can be found by $n_{eff} = 1 - \frac{1}{2}\left(\frac{u_{\mu\nu}\lambda}{\pi D_{in}}\right)^2$,

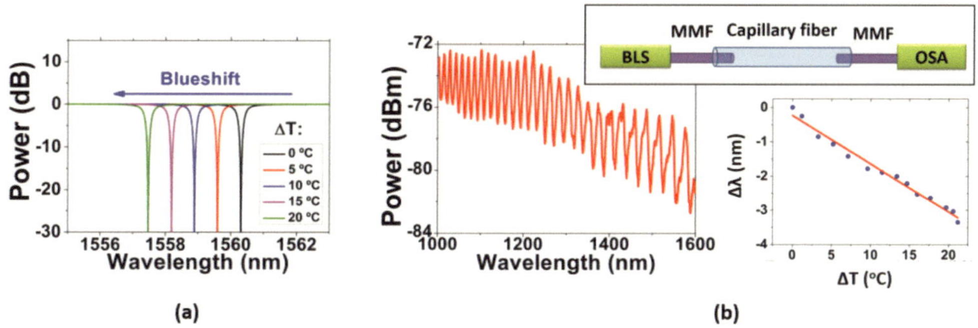

Figure 8.
(a) Simulated transmittance of a capillary fiber 2 cm long and with 160 μm inner diameter and 240 μm outer diameter. (b) Experimental transmitted spectrum for a 13 cm long PMMA capillary fiber. Insets stands for the experimental setup (BLS: broadband light source; MMF: multimode fiber; OSA: optical spectrum analyzer) and for the wavelength shift (Δλ) as a function of the temperature variation (ΔT).

where D_{in} is inner diameter of the capillary and $u_{\mu\upsilon}$ is a root of the equation $J_{\upsilon-1}(u_{\mu\upsilon}) = 0$, where J is the Bessel function [34, 35].

$$P_{out} = P_{in} \exp\left\{\left(\frac{L}{D_{in}\tan\theta_1}\right)\ln\left[1 - \frac{\left(1-\Gamma^2\right)^2}{\left(1-\Gamma^2\right)^2 + 4\Gamma^2 \sin^2\left(\frac{2\pi n_2 d}{\lambda}\sqrt{1-\left(\frac{n_1}{n_2}\right)^2 \sin^2\theta_1}\right)}\right]\right\} \tag{10}$$

$$\Gamma = \frac{\sqrt{1-\sin^2\theta_1} - \frac{n_2}{n_1}\sqrt{1-\left(\frac{n_1}{n_2}\right)^2 \sin^2\theta_1}}{\sqrt{1-\sin^2\theta_1} + \frac{n_2}{n_1}\sqrt{1-\left(\frac{n_1}{n_2}\right)^2 \sin^2\theta_1}} \tag{11}$$

Figure 8a presents the simulation for a transmission minimum of a PMMA capillary with 2 cm length, inner diameter 160 μm, and outer diameter 240 μm for different temperature conditions. In the simulation, both thermal expansion and thermo-optic effects are considered. It is seen that, when the temperature increases, the transmission dip is expected to blueshift. It is worth saying that, for the considered capillary fiber, the thermo-optic effect is the dominant effect.

To experimentally test the proposed sensor, a PMMA capillary fiber was fabricated (with inner diameter 160 μm and outer diameter 240 μm) and its performance as a temperature sensor was measured. In the experimental setup, light from a broadband light source was coupled to the capillary fiber (13 cm length) and collected from it by using silica multimode fibers, as shown in the inset of **Figure 8b**. A typical transmission spectrum is shown in **Figure 8b**, and the wavelength shift (Δλ) as a function of the temperature variation (ΔT) is shown in the inset in **Figure 8b**. A sensitivity of (-140 ± 6) pm/°C was measured. This sensitivity is around 14 times higher than conventional Bragg gratings-based temperature sensors [36].

8. Conclusions

In this chapter, we presented the recent research yields of our group in the State University of Campinas (Unicamp, Brazil) regarding ultra-simplified microstructured optical fibers designs for sensing applications. In this context, we firstly discussed hollow and metal-filled embedded-core fibers as pressure and temperature sensors. We showed that the achieved sensitivities with this structure are high when compared to other sensors which employ much more sophisticated fiber designs. Additionally, we described surface-core fibers as refractive index and directional curvature sensors by employing fiber Bragg gratings and a plasmonic configuration. Finally, polymer capillary fibers were presented as an even simpler structure for temperature sensing.

The results presented herein demonstrate the great potential of capillary-like fibers to act as sensors for multiple purposes. Hence, it is possible to identify that this minimalist approach for the design of microstructured fiber sensors consists of a novel and very promising avenue for obtaining sensors with simplified structures and optimized performances.

Acknowledgements

We thank M. A. R. Franco, V. A. Serrão, G. Chesini, S. Aristilde, R. Oliveira, T. H. R. Marques, A. Matikainen, and H. Ludvigsen for their contributions to the development of the research reported herein.

This work was funded by the São Paulo Research Foundation (Fapesp, grant #2014/50632-6), the Financiadora de Estudos e Projetos (Finep, grant #01.12.0393.00), and the National Council for Scientific and Technological Development (CNPq).

Author details

Jonas H. Osório* and Cristiano M. B. Cordeiro
"Gleb Wataghin" Physics Institute, State University of Campinas, Brazil

*Address all correspondence to: cmbc@ifi.unicamp.br

References

[1] Anuszkiewicz A, Statkiewicz-Barabach G, Borsukowski T, Olszewski J, Martykien T, Urbanczyk W, et al. Sensing characteristics of the rocking filters in microstructured fiber optimized for hydrostatic pressure measurements. Optics Express. 2012; **20**(21):23320-23330

[2] Osório JH, Hayashi JG, Espinel YAV, Franco MAR, Andrés MV, Cordeiro CMB. Photonic-crystal fiber-based pressure sensor for dual environment monitoring. Applied Optics. 2014;**53**: 3668-3672

[3] Kim BH, Lee SH, Lin A, Lee A, Lee J, Han W. Large temperature sensitivity of Sagnac loop interferometer based on the birefringent holey fiber filled with metal indium. Optics Express. 2009;**17**: 1789-1794

[4] Reyes-Vera E, Cordeiro CMB, Torres P. Highly sensitive temperature sensor using a sagnac loop interferometer based on a side-hole photonic crystal fiber filled with metal. Applied Optics. 2017;**56**(2):156-162

[5] Hautakorpi M, Mattinen M, Ludvigsen H. Surface-plasmon-resonance sensor based on three-hole microstructured optical fiber. Optics Express. 2008;**16**:8427-8432

[6] Doherty B, Thiele M, Warren-Smith S, Schartner E, Ebendorff-Heidepriem H, Fritzsche W, et al. Plasmonic nanoparticle-functionalized exposed-core fiber – an optofluidic refractive index sensing platform. Optics Letters. 2017;**41**(21):4395-4398

[7] Osório JH, Chesini G, Serrão VA, Franco MAR, Cordeiro CMB. Simplifying the design of microstructured optical fibre pressure sensors. Scientific Reports. 2017;**7**:2990

[8] Chesini G, Osório JH, Serrão VA, Franco MAR, Cordeiro CMB. Metal-filled embedded-core capillary fibers as highly sensitive temperature sensors. IEEE Sensors Letters. 2018;**2**(2)

[9] Osório JH, Oliveira R, Aristilde S, Chesini G, Franco MAR, Nogueira RN, et al. Bragg gratings in surface-core fibers: refractive index and directional curvature sensing. Optical Fiber Technology. 2017;**34**:86-90

[10] Matikainen A, Osório JH, Nyman M, Juntunen T, Cordeiro CMB, Ludvigsen H. Plasmonic surface-core fibers for high sensitivity refractive index sensing. In preparation

[11] Osório JH, Marques THR, Figueredo IC, Serrão VA, Franco MAR, Cordeiro CMB. Optical sensing with antiresonant capillary fibers. In: 25th International Conference on Optical Fiber Sensors, Proc. SPIE 10323. 2017. p. 103233U

[12] Rugeland P, Sterner C, Margulis W. Visible light guidance in silica capillaries by antiresonant reflection. Optics Express. 2013;**21**:29217-29222

[13] Szpulak M, Martynkien T, Urbanczyk W. Effects of hydrostatic pressure on phase and group birefringence in microstructured holey fibers. Applied Optics. 2004;**43**: 4739-4744

[14] Primak W, Post D. Photoelastic constants of vitreous silica and its elastic coefficient of refractive index. Journal of Applied Physics. 1959;**30**:779

[15] Timoshenko S, Goodter JN. Theory of Elasticity. New York: McGraw-Hill; 1969

[16] Osório JH, Cordeiro CMB. Optical sensor based on two in-series

birefringent optical fibers. Applied Optics. 2013;**52**:4915-4921

[17] Fu HY, Tam HY, Shao L, Dong X, Wai PKA, Lu C, et al. Pressure sensors realized with polarization-maintaining photonic crystal fiber-based Sagnac interferometer. Applied Optics. 2008; **47**(15):2835-2839

[18] Martynkien T, Szpulak M, Statkiewicz G, Golojuch G, Olszewski J, Urbanczyk W, et al. Measurements of sensitivity to hydrostatic pressure and temperature in highly birefringent photonic crystal fibers. Optical and Quantum Eletronics. 2007;**39**:481-489

[19] Martynkien T, Statkiewicz-Barabach G, Olszewski J, Wojcik J, Mergo P, Geernaert T, et al. Highly birefringent microstructured fibers with enhanced sensitivity to hydrostatic pressure. Optics Express. 2010;**18**(14): 15113-15121

[20] Xin Y, Dong X, Meng Q, Qi F, Zhao C. Alcohol-filled sidehole fiber Sagnac interferometer for temperature measurement. Sensors and Actuators A: Physical. 2013;**193**:182-185

[21] Naeem BH, Kim B, Kim YC. High-sensitivity temperature sensor based on a selectively-polymer-filled two-core photonic crystal fiber in-line interferometer. IEEE Sensors Journal. 2015;**15**(7):3998-4003

[22] Liang W, Huang Y, Xu Y, Lee RK, Yariv A. Highly sensitive fiber Bragg gratings refractive index sensors. Applied Physics Letters. 2005;**86**:151122

[23] Iadicicco I, Cusano A, Cutolo A, Bernini R, Giordano M. Thinned fiber Bragg gratings as high sensitivity refractive index sensor. Photonics Technology Letters. 2004;**16**(4): 1149-1151

[24] Esteban O, Díaz-Herrera N, Navarrete M, González-Cano A. Surface-plasmon resonance sensors based on uniform-waist tapered fibers in a reflective configuration. Applied Optics. 2006;**45**:7294-7298

[25] Hautakorpi M, Mattinen M, Ludvigsen H. Surface-plasmon-resonance sensor based on three-hole microstructured fiber. Optics Express. 2008;**16**:8427-8432

[26] Hibbeler RC. Mechanics of Materials. 8th ed. Boston: Pearson Prentice Hall; 2011. 978-0-13-602230-5

[27] Kreuzer M. Strain Measurement with Fiber Bragg Gratings Sensors. Darmstadt, Germany: HBM; 2006

[28] Gander MG, Mac Pherson WN, McBride R, Jones JDC, Zhang L, Bennion I, et al. Bend measurement using Bragg gratings in multicore fibre. Electronics Letters. 2000;**36**(2):120-121

[29] Flockhart GMH, Mac Pherson WN, Barton JS, Jones JDC. Two-axis bend measurement with Bragg gratings in multicore optical fiber. Optics Letters. 2003;**28**(6):387-389

[30] Araújo FA, Ferreira LA, Santos JL. Simultaneous determination of curvature, plane of curvature, and temperature by use of a miniaturized sensing head based on fiber Bragg gratings. Applied Optics. 2002;**41**(13): 2401-2407

[31] Chen X, Zhang C, Webb DJ, Peng G, Kalli K. Bragg grating in a polymer optical fiber for strain, bend and temperature sensing. Measurement Science and Technology. 2010;**21**: 094005

[32] Villatoro J, Newkirk AV, Antonio-Lopez E, Zubia J, Shülzgen A, Amezcua-Correa R. Ultrasensitive vector bending

sensor based on multicore optical fiber. Optics Letters. 2016;**41**:832-835

[33] Guzman-Sepulveda JR, May-Arrioja DA. In-fiber directional coupler for high-sensitivity curvature measurement. Optics Express. 2013;**21**: 11853-11861

[34] Lai C, You B, Lu J, Liu T, Peng J, Sun C, et al. Modal characteristics of antiresonant reflectıng pıpe waveguides for terahertz waveguiding. Optics Express. 2010;**18**(1)

[35] Marcatili EAJ, Schmeltzer RA. Hollow metallic and dielectric waveguides for long distance optical transmission and lasers. The Bell System Technical Journal. 1964;**43**(4)

[36] Oliveira R, Osório JH, Aristilde S, Bilro L, Nogueira RN, Cordeiro CMB. Simultaneous measurement of strain, temperature and refractive index based on multimode interference, fiber tapering and fiber Bragg gratings. Measurement Science and Technology. 2016;**27**:075107

Chapter 3

Fiber Bragg Gratings as e-Health Enablers: An Overview for Gait Analysis Applications

Maria de Fátima Domingues, Cátia Tavares, Tiago Leite, Nélia Alberto, Cátia Leitão, Carlos Marques, Ayman Radwan, Eduardo Rocon, Paulo Antunes and Paulo André

Abstract

Nowadays, the fast advances in sensing technologies and ubiquitous wireless networking are reflected in medical practice. It provides new healthcare advantages under the scope of e-Health applications, enhancing life quality of citizens. The increase of life expectancy of current population comes with its challenges and growing health risks, which include locomotive problems. Such impairments and its rehabilitation require a close monitoring and continuous evaluation, which add financial burdens on an already overloaded healthcare system. Analysis of body movements and gait pattern can help in the rehabilitation of such problems. These monitoring systems should be noninvasive and comfortable, in order to not jeopardize the mobility and the day-to-day activities of citizens. The use of fiber Bragg gratings (FBGs) as e-Health enablers has presented itself as a new topic to be investigated, exploiting the FBGs' advantages over its electronic counterparts. Although gait analysis has been widely assessed, the use of FBGs in biomechanics and rehabilitation is recent, with a wide field of applications. This chapter provides a review of the application of FBGs for gait analysis monitoring, namely its use in topics such as the monitoring of plantar pressure, angle, and torsion and its integration in rehabilitation exoskeletons and for prosthetic control.

Keywords: fiber Bragg gratings, e-Health enablers, gait analysis, plantar pressure, foot shear pressure, gait joint monitoring, instrumentation of prosthetic limbs

1. Fiber Bragg gratings: an introduction

Fiber Bragg gratings (FBGs) are sensing elements based on the longitudinal modulation of the refractive index of the optical fiber core. This type of device has all the advantages associated with optical fiber sensors, with the added feature of easily multiplexing several sensing points along one single fiber.

The production methodology of FBGs has evolved significantly since its initial approach. In the late 1970s, it was shown that optical fibers can be photosensitive, opening the door for FBGs production and its applications, both as sensing devices and in optical communications [1]. In 1981, Lam and Garside suggested that the formation of the FBGs was related to the interaction between UV light with defects in the doped silica core. Such findings lead to the later confirmation that the refractive index changes could be induced by doping the optical fibers core with germanium, given a new insight on the FBGs production [2, 3].

One decade has passed since new breakthroughs emerged regarding the FBGs production methodology. In 1989, Meltz et al. reported an FBG external inscription technique. The authors used a split 244 nm beam, which was later recombined in order to produce an interference pattern in the optical fiber core [4, 5]. With this technique, the authors were able to create a periodic and permanent change in the optical fiber core refractive index [5]. The reflected Bragg wavelength can be adjusted by changing the angle between the two split beams. In that way, the period of the interference pattern and the refractive index will change accordingly.

Alternatively, FBGs can be inscribed using phase masks, which are periodic patterns usually etched onto fused silica. In this technique, when the radiation from a UV laser is incident in the phase mask, the diffracted orders +1 and −1 are maximized, while the remaining ones are suppressed, creating an interferometric reflective pattern along the optical fiber core [6]. In **Figure 1**, the FBG inscription based on the phase mask technique as well as a representation of the FBG sensing mechanism is shown.

The FBG operational principle consists in monitoring the Bragg wavelength (λ_{Bragg}) shift reflected by the grating, as a function of the monitored parameter. The Bragg wavelength is dependent on the effective refractive index of the fiber core (n_{eff}) and the grating period (Λ) by the relation [4]:

$$\lambda_{Bragg} = 2 n_{eff} \Lambda \quad (1)$$

Therefore, the Bragg wavelength can be actuated by variations in the grating period or in optical fiber core effective refractive index. So, the Bragg wavelength dependence on strain and temperature can be translated by:

$$\Delta\lambda_{Bragg} = \lambda_{Bragg}(1 - \rho\alpha)\Delta\varepsilon + \lambda_{Bragg}(\alpha + \xi)\Delta T, \quad (2)$$

where the first term refers to the strain influence on the λ_{Bragg} and the second describes the temperature effect. Hence, in Eq. (2), $\Delta\lambda_{Bragg}$ represents the shift of the Bragg wavelength, while ρ, α, and ξ are the photoelastic, thermal expansion,

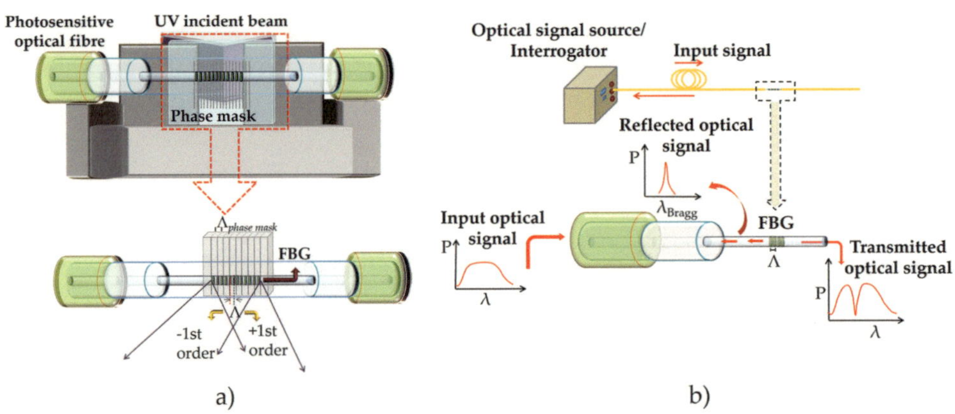

Figure 1.
(a) Schematic representation of the setup typically used to inscribe FBG sensors in photosensitive optical fiber using the phase mask methodology; and (b) working principle of an FBG sensor.

and thermo-optic coefficients of the fiber, respectively; Δε and ΔT corresponds to strain and temperature variations.

The FBG sensing mechanism comprises of a spectral broadband optical signal launched into the fiber, and an optical spectra analyzer to monitor the Bragg wavelength shifts. At the grating region, the Bragg wavelength component of the spectrum will be reflected, while in the transmitted optical signal that same Bragg wavelength component will be missing, as illustrated in **Figure 1b**.

Based in the described mechanisms, FBG sensors have a wide field of applications that range from their use for structural health monitoring, in oil and aeronautic industry and also as biomedical sensors and e-Health enablers, among others. Moreover, as the FBGs are elements with only few millimeters long, several gratings can be inscribed along the same optical fiber, allowing to multiplex a diverse network of sensing elements.

2. Gait analysis: relevance and impact in an e-Health scenario

Gait analysis research was given a pilot role in the nineteenth century, when the study of gait parameters started to be relevant in sports and medicine [7]. Regarding the medical point of view, from gait pattern analysis, a change in its normal parameters can reveal key information on patient's quality of life and/or in the evolution of different diseases. Gait disorders affect a large number of world population, since they are direct consequence of neurodegenerative diseases, such as spinal amyotrophic, multiple sclerosis, amyotrophic lateral sclerosis, neuromuscular diseases, cerebrovascular and cardiovascular pathologies, or even the physiological aging process [8–12]. Neurodegenerative diseases can be reflected in gait by showing a poor balance, a slower pace, shorter steps, lower free speed, and higher cadence [8–11].

The study of dynamic characteristics of human gait for clinical purposes has been widely reported lately. It aims to enhance the life's quality of patients suffering from gait disorders, and also, for their early detection, to enable early diagnosis and an adaptable treatment according to the evolution of the diseases or disorders [7, 13–16].

2.1 Gait analysis: gait cycle pattern

Gait analysis can be seen as the comprehensive study of the human locomotion, which as previously mentioned, has a major role in physical rehabilitation assessment, disorder diagnosis, surgical decision, and recovering follow up. Such study comprises the kinematic analysis (joint angles, angular velocities, and accelerations) and the kinetic analysis (ground reaction and joint forces) during the gait cycles [17, 18].

One gait cycle is the period of time between two consecutive contacts of the heel of the same foot with the floor. Generally, a cycle can be divided in two major phases: the stance phase, corresponding to the period in contact with the ground, which lasts for ~60% of the cycle; and the swing phase, corresponding to the period when there is no contact with the floor, and has a duration of ~40% of the total gait cycle [12, 19]. In **Figure 2**, the different phases are illustrated, along with events and periods that characterize a gait cycle.

The gait cycle can be further subdivided into six periods and eight functional events, five during the stance phase and three in the swing phase. Considering only one limb, the stance phase encompasses three different support periods. The first consists in a period of a double support, which is followed by single

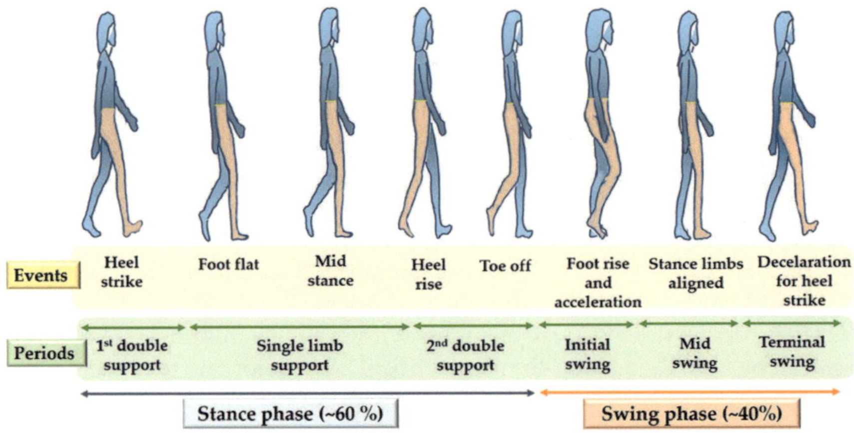

Figure 2.
Representation of the stance and swing phases of a gait cycle.

support and ends with the second double support period [18–20]. The double support period corresponds to the percentage of the cycle when both feet are simultaneously in contact with the floor and it describes the smooth transition between the left and the right single limbs support [18]. During the first double support, the heel strikes the floor (heel strike), marking the beginning of the gait cycle. The cycle evolves then toward the single period support, with the foot moving down toward the floor into a foot-flat position, where a stable support base is created for the rest of the body. Within the single support phase, the body is propelled over the foot, with the hip joint vertically aligned with ankle joint in the event characterized as the mid stance. From that point onward, the second double support phase starts, with the lower limb moving the body center of mass forward during the heel rise event, where the heel loses contact with the floor. The last contact of the foot with the floor is made by the big toe (hallux), at the toe off event, which also marks the end of the stance phase and the beginning of the swing phase [20].

During the swing phase, there is no contact between the plantar foot and the floor, and the limb continues its movement forward, which can be divided into three different periods: initial swing, mid swing, and terminal swing. In the initial swing, the lower limb vertical length should be reduced, for the foot to clear the floor and to accelerate forward by flexing the hip and knee, together with ankle dorsiflexion. The mid swing is characterized by the alignment of the accelerating limb with the stance limb. In this phase, the ankle and the hip joints are aligned. During the terminal swing, the limb undergoes a deceleration while it prepares for the contact with the floor, in the heel strike of the start of a new cycle [19–21]. As described, the swing phase is characterized by accelerations and decelerations of the lower limb, which require a more demanding muscular effort at the hip level [18].

2.2 Gait parameters

Gait analysis is a systematic procedure that allows the detection of negative deviations from normal gait pattern, as well as their causes. Based on such analysis, it is possible to quantify the parameters involved in the movement of the lower limbs and retrieve the mechanisms that rule the human body movement [22]. Based on the gait cycle pattern described earlier, there are several parameters that can be physically monitored in order to assess the patient's health: anthropometric, spatio-temporal, kinematic, kinetic, and dynamic electromyography (EMG), as shown in **Table 1** [22]. From such parameters, the ones that require a more specialized technology to be monitored outside the clinical environment, and therefore passible of being monitored in a gait e-Health architecture, are [7, 23, 24]:

- the stance and the swing phases duration for each foot;
- the walking velocity and gait cadence (number of steps per unit of time);
- The step length, width (distance between two equivalent points of both feet), and angles (direction of the foot during gait);
- the body posture (bending and symmetry) and the existence of tremors;
- the shear and the foot plantar pressure during the stance phase; and
- the direction and alignment of the limb segments with the ankle, knee, and hip joints.

The act of walking implies the movement of the whole body, and specifically, it requires a synchronized movement of each lower limb apart. Therefore, the gait pattern of an individual can be affected by a disorder in any segment of the body, like for instance, problems in the spinal cord or from a reduced knee flexion in patients with an anterior cruciate ligament reconstruction [25]. For that reason, the analysis of the gait cycle is a vital tool for the biomechanical mobility monitoring, as it can give crucial information not only about the lower limbs health condition, but also allows to infer details about other possible pathologies related to the dynamic movement of the body [26]. So, by monitoring the parameters previously listed, it is possible to assess the health conditions for the body parts involved in walking, namely the lower limbs and its joint. These parameters can be analyzed using objective and subjective techniques [7, 27, 28].

The subjective analysis is based on the observation of the patient while walking, and is generally performed in clinical environment under the supervision

Parameters	Definition	Evaluation of:
Anthropometric	Related to the corporal dimensions of the human body.	Age, gender, height, weight, limb length, and body mass index.
Spatio-temporal	General gait parameters used for a simple objective gait evaluation, considering the time-distance characteristics.	Step and stride length, step width, cadence, velocity, stance and swing phases, and gait cycle events (for instance, heel strike).
Kinematic	Quantification of movements and geometric description of the lower limbs motion, without reference to forces.	Joint and segments angles, angular motion, acceleration, and segment trajectory.
Kinetic	Evaluation of the forces involved in the body locomotion.	Ground reaction forces, torque, and momentum.
EMG	Refers to the analysis of muscular activity, generally performed by using EMG surface electrodes.	Motor unit action parameters.

Table 1.
Parameters generally used for gait analysis (adapted from [22]).

of a doctor or a therapist. For this analysis, the patient is asked to vary several gait-related parameters, while walking in a predetermined circuit [7]. This type of analysis is bit limited in the information that can be retrieved, nevertheless, this could be useful for an initial evaluation and posterior decision on which objective techniques should be used. In contrast to the subjective techniques, objective gait analysis is more of a quantitative evaluation of the parameters listed above. This type of analysis requires the use of different types of equipment and procedures to measure the gait parameters. These methodologies can be categorized according to the technology used, varying from the ones based on imaging, instrumented walking platforms or floor sensors, and wearable sensors [7, 24, 29].

For an e-Health architecture, the most suitable technology would be the one built using wearable sensors, which would be able to acquire the patient gait parameters, everywhere and under any conditions. Among those, FBGs can be considered as an objective technique for gait analysis (allows the quantification of parameters during gait analysis), which could be used in instrumented platforms or as wearable sensors [30]. Recently, the use of FBGs as wearable sensors for remote monitoring of patients has been reported [12, 31, 32].

2.3 Gait pattern monitoring: e-Health architecture

The Internet of Things (IoT) concept is the fusion between pervasive network connectivity and the computing capability expanded to sensing devices and objects, able to acquire and exchange data autonomously. In recent years, due to the potential gains brought to the citizens' quality of life, IoT is seen as a whole platform able to bridge people and objects by integrating the smart concept into people's life, namely, smart cities, homes, wearables, and mobility [33].

Within the vast field of applications provided by IoT, e-Health stands out as one of the most influential topics on life-quality of humans, as smart and connected healthcare services have been requested more enthusiastically. e-Health is gaining too much attention mainly due to the joint effect of the increase of insufficient and ineffective healthcare services, allied to the change in population demographics and the increasing demand of such change entails. The world's population aged over 60 years is expected to reach 2 billion by 2050 [34], which implies the rise of chronic diseases that may be translated on different degrees of mobility impairments, requiring a close monitoring and a patient-centered healthcare service, where the healthcare providers and patients are pervasively connected [12, 33]. Also aligned with such demands, the market for home medical devices is set to significantly grow from $27.8 billion in 2015 up to nearly $44.3 billion by 2020 [35]. The increase in available e-Health solutions is a remarkable step toward improving the healthcare services, along with the autonomy of debilitated or impaired citizens. Fundamentally, e-Health can be seen as the solution to help the elders and patients with chronic illness to live an active life, without compromising their mobility or daily routine [32].

e-Health systems use remote monitoring architectures composed of sensing devices responsible to collect patients' physiological information, analyze and store such data in the cloud. The information can afterward be wirelessly sent to the healthcare professionals for a decision/action. The continuous flow of information on the patient's condition improves the provided service at a lower cost, while simultaneously enhancing the life quality of patients, who need continuous attention [33].

The patients' physiological information can be collected by networked sensors, integrated in smart wearable systems, or placed within the patients living environment. Considering the specific case of gait analysis, the continuous, automatic, and remote monitoring of gait of impaired or under rehabilitation citizens allows the objective assessment for preventive and proactive supervision of the pathologies, as well as to closely assist the therapies in progress [36]. In this scenario, wearable sensing architecture allows not only the evaluation of the patients in the course of daily life activities, but also provides the feedback on the recovering/rehabilitation therapy to patients and medical staff through ubiquitous connectivity. Based on that feedback, new therapeutic instructions can be given remotely to the patient, maximizing the efficiency of the provided healthcare services [31, 32].

An e-Health architecture to monitor the gait pattern of a citizen/patient comprises three key elements: the monitoring system composed of a sensors' network (preferably wearable); a computer/analysis system to collect, analyze, and store the data; and finally, the wireless mobile gateway, responsible for data processing and wireless transmission to medical servers and decision centers [31, 32]. **Figure 3** schematizes the typical architecture involved in an e-Health scenario.

The first part of the considered architecture is responsible for the data sensing and consequently, for the information given to action centers. Therefore, it is crucial that the sensing network is as accurate and reliable as possible. The use of FBGs as IoT and e-Health enablers is becoming increasingly common, due to their sensing characteristics, when compared with the ones of their electronic counterparts, namely small size (in the order of micrometers), biocompatibility, multiplexing capability, immunity to electromagnetic interference, in addition to their high accuracy and sensitivity, even for applications in challenging environments [37–39]. Consequently, FBGs can be used as a reliable solution for the integration in e-Health architectures, as for monitoring sensing systems in biomechanics and physical rehabilitation. Some examples can be mentioned, covering the detection of bone strains, mapping of gait plantar and shear pressures, measuring of pressures in orthopedic joints and angles between the body segments, as have already been successfully reported [12, 30].

In the following sections, the use of FBGs to monitor different body segments involved in gait will be explored.

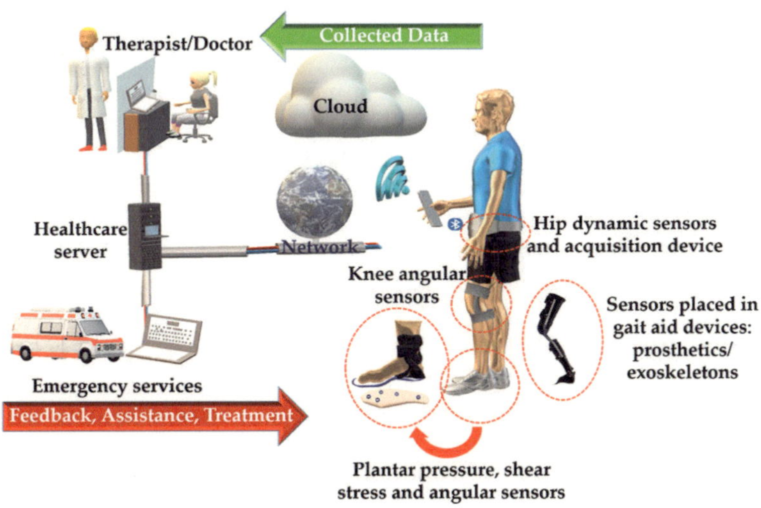

Figure 3.
Possible architecture for a gait e-Health monitoring system.

3. Plantar pressure and shear analysis

The assessment of plantar and shear pressures is of great importance for the gait health evaluation analysis, aiming to understand the effects induced in/by the body and to prevent the ulceration of the foot [40]. The pressure ulcers occur when tissue is compressed during prolonged periods of time, resulting in a wound that can infect and cause amputation, or in more severe cases, the patient's death [41]. An early identification of individuals at risk of foot ulceration (people with diabetes mellitus and peripheral neuropathy) is one of the primary means to reduce its incidence [40, 41]. Due to the poor load distribution, resulting from the reduced sensitivity of the foot, abnormally high plantar pressures occur in certain areas of the foot, and when that happens, it can lead to the growth of pressure sores in these locations [42]. The most affected areas are those with bony prominences, such as under the metatarsal bones, where the majority of plantar neuropathic ulcers occur [43]. Correct and continuous mapping of plantar pressure can prevent the occurrence of these pathologies, with the adoption of different walking habits or the use of correction equipment. On the other hand, in the case of the existence of ulcers, a redistribution of the forces imposed on the foot during walking aids healing and prevents further ulceration.

The force applied to the skin surface by a supporting structure has two components: the pressure acting normal to the surface and the shear stress acting in a tangential direction. Many authors have suggested that shear stresses have a pathogenic factor in the development of plantar ulcers [40–43]. This shear stress exists if there is sliding between two surfaces (foot and shoe), and it is closely related to friction [44]. Despite the importance of shear monitoring in assessing gait patterns, only normal pressure is widely reported. The lack of a validated and commercially available shear stress sensor is one of the main reasons why shear analysis is not as referenced as plantar pressure.

There are several solutions in the market for measuring plantar pressure, static in the form of fixed platforms, and wearable as shoes insoles. In regards of the importance of gait-related pathologies in the general population, and in the elder generation in particular, several works have been developed to improve the state-of-the-art. The literature reports the use of various technologies of plantar pressure and shear sensors, such as magneto-resistors, strain gauges, piezoelectric materials, capacitive sensors, and micro-strip antennas and coils.

As an alternative to these electronic devices, optical fiber-based sensors stand out due to their small diameter (hundreds of micrometers) and robustness, biocompatibility, high precision, electromagnetic insensitivity, as well as being electrically free at the point of measurement and owning the property of being able to multiplex several sensors in the same fiber, which allows to simultaneously monitor different parameters [45, 46].

Following such path, research studies have been carried out with FBG-based sensors used to measure not only plantar pressure but also shear parameters.

3.1 Plantar pressure sensors

The plantar pressure monitoring devices can be presented as fixed platforms or as insoles to be used directly in the footwear. Platform systems are typically constructed of several pressure sensors arranged in an array embedded in the floor or in a rigid platform. These systems can be used for static and dynamic studies, but are generally restricted to clinics and laboratories. In the case of the static tests, the patient stands still on the platform. On the other hand, for the dynamic tests, the platform is placed on the floor and the patient walks through it.

The application of these types of measuring systems has the advantage of being easy to use, since the platforms are flat and stationary. Nevertheless, these systems also present disadvantages, since they are influential to the patient's gait, once during the examination, he/she will have to tread specific areas of the platform surface [47].

Insole sensors can be incorporated into shoes, so that the measurements truly reflect the interface between the foot and the shoe. These systems, as they are flexible and portable, allow a greater accuracy of the acquired data, regarding the natural gait of the patient, and also greater variety of studies with different walking tasks, footwear, and even diverse floors/terrains [47]. However, insoles usually have a reduced number of sensors compared to platforms. The main requirements for the development of wearable/in-shoe sensors are: mobility, reduced number of cables, low power consumption, low cost, high acquisition frequency, proper sensitivity, noninvasiveness, and do not represent any danger to the user. Therefore, fiber optic sensors, due to their characteristics, have proven to be a reliable solution in this type of applications. Also, within the range of fiber optic sensors, the FBGs seem to be the best solution, since their multiplexing capability allows to have multiple sensors into a single fiber, reducing the number of cables needed in the insole. In this section, some recently developed work using optical fibers with Bragg gratings, as plantar pressure sensors in fixed platforms and in-shoe equipment, will be described.

The first work with FBG sensors incorporated in platforms for the measurement of plantar pressure was published in 2003, when Hao's team developed an insole shaped device with a silica optical fiber with five FBGs [48]. The insole was constituted of 10 layers of carbo-epoxy, among which the optical fiber was placed. The sensing FBG units were placed strategically at the main pressure points (heel and metatarsal areas). The device was tested in static tests to determine which areas have the greatest and lowest pressure at different user positions. The results showed that the sensors had an average sensitivity of 5.44 pm/N.

In 2014, Suresh et al. published a work comparing the use of FBGs and piezoelectric (PZT) sensors for gait monitoring at low and high speeds [49]. To manufacture the optical sensing platform, the FBGs were embedded between layers of a carbon composite material (CCM) in the form of an arc. After that, both types of sensors were placed on the underside of a commercial shoe (**Figure 4a**). For the dynamic test and to verify the behavior of both types of sensors, a healthy male walked on a treadmill wearing those shoes at various speeds. For the FBGs sensors, a mean pressure sensitivity of 1.3 pm/kPa was obtained. The study revealed that the FBG sensors have a better performance in the static moments and at lower speeds, while the piezoelectric sensors had greater performance for higher speeds.

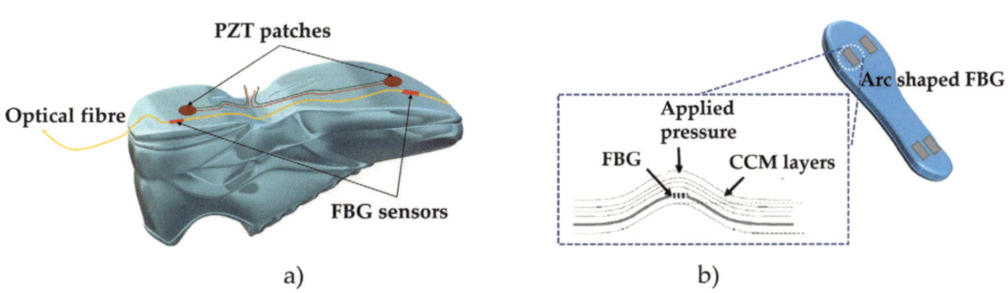

Figure 4.
(a) Scheme of the shoe with the attached FBG and PZT sensors (adapted from [49]); and (b) schematic representation of an arc shaped FBG pressure sensor (left) and the insole sensing scheme (right) (adapted from [50]).

Other approach was made by the same team, in which similar FBG cells were incorporated in an insole structure, as can be verified in the **Figure 4b**. The device was constituted by four arc-shaped cells, strategically placed in the forefoot and heel area. In this study, the plantar pressure was analyzed in both the fixed platform and the in-shoe systems. An average pressure sensitivity of 1.2 pm/kPa was obtained [50].

In 2016, Liang et al. proposed a sensing system based on six FBGs inscribed into a single fiber, which was embedded in silicone rubber [51]. The data registered by the optical sensors were compared to the ones collected through an i-Step P1000 digital pressure plate with 1024 barometric sensors. For the sensors' validation, 11 participants were tested, and according to the results, the viability of the optical sensor for this kind of measurement was demonstrated. Additionally, four different foot supporting types were successfully identified.

In 2017, Domingues et al. reported the development of two noninvasive solutions with FBGs in silica optical fiber incorporated in cork to monitor the body center of mass displacements and vertical ground reaction forces induced in the foot plantar surface during gait [12, 32]. One of the solutions, containing five FBGs, was developed to act as a fixed platform, and the other, with six FBG sensors, to be used as an instrumented insole to be adapted in a shoe as shown in **Figure 5**. Although the insole is made of five FBGs multiplexed in the same fiber, a clear isolation of each sensing point was also demonstrated, as seen in **Figure 6a**. Upon the calibration of the sensors located at point 1 (heel area), when the increasing load is applied in that point, only the FBG 1 shows a Bragg wavelength shift, proportional to the load applied (**Figure 6b**) [12].

The obtained results demonstrated the accuracy and reliability of the proposed systems to monitor and map the vertical active forces on the foot's plantar area during gait, **Figure 7**, with a sensitivity up to 11.06 pm/N.

The top graphic representation corresponds to the values independently registered by the five FBGs along time during two gait cycles, where the sum of the forces acquired by each FBG corresponds to the typical gait pattern [12]. In more detail, in the bottom graphic representation, it is possible to see which points of the insole are more actively pressed during the stance phase of the gait cycle. The dark blue representations in the foot, corresponds to the foot area that supports a higher load in the different stages of the stance phase [12, 32].

In the same year, a polymer optical fiber (POF) sensing system based on FBGs to measure foot plantar pressure was also described [52]. The plantar pressure signals were detected by five FBGs recorded in a cyclic transparent optical polymer (CYTOP) fiber, which was embedded in a cork platform in the form of an insole to monitor plantar pressure during gait. Initially, two studies were made with the insole as a fixed platform, one in which the user walked through the sensing

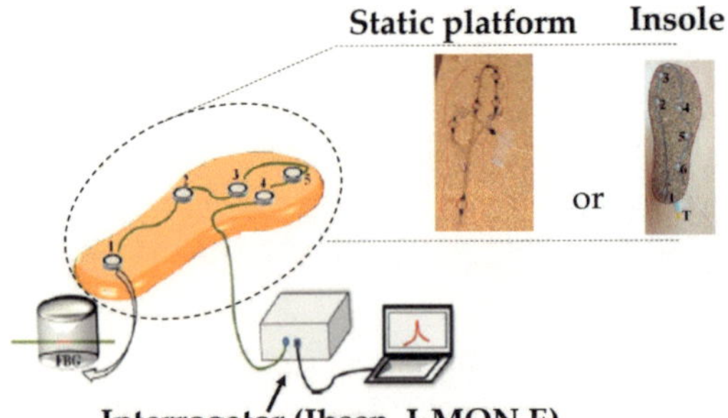

Figure 5.
Schematic representation of the cork insole FBG monitoring system (adapted from [12]).

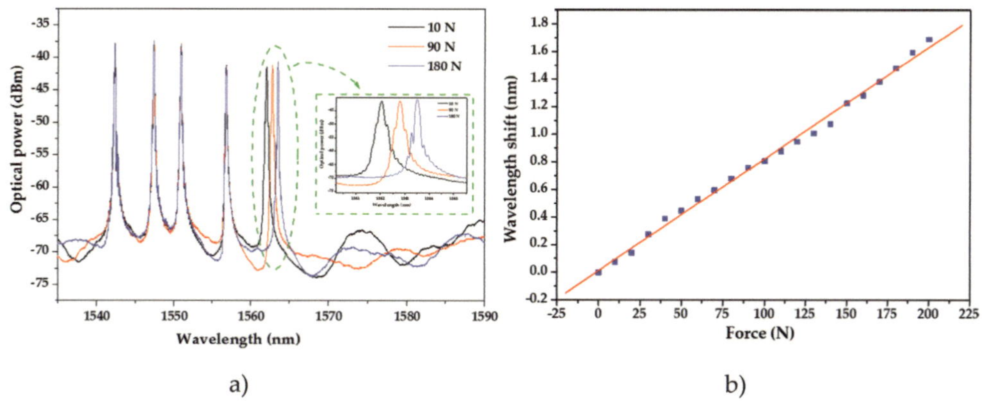

Figure 6.
(a) Reflection spectra of the five FBGs multiplexed in the cork insole for three different load values; and (b) Bragg wavelength shift dependence on the load applied for FBG1 (adapted from [12]).

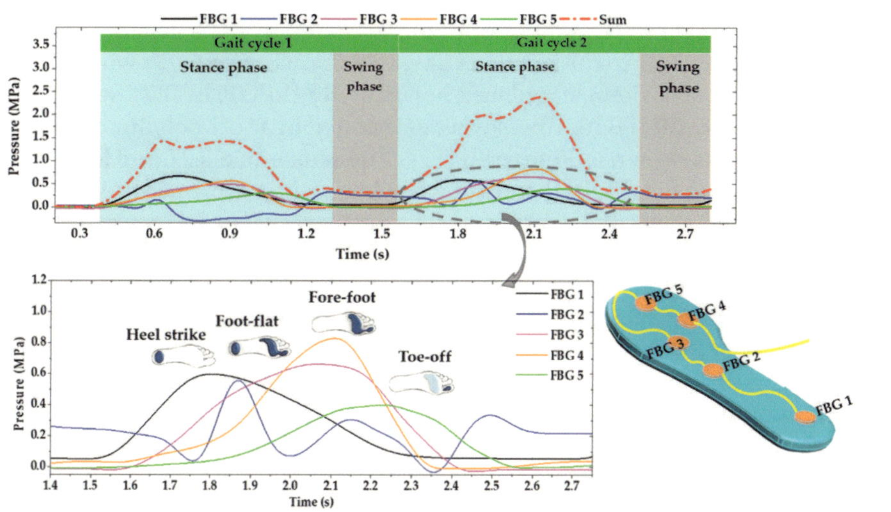

Figure 7.
Representation of two complete gait cycles registered using a cork insole instrumented with five multiplexed FBGs (adapted from [12, 32]).

structure and another where he stood in the platform just moving the body center of mass. The data obtained from this device in both tests showed good repeatability and a sensitivity twice as high as the solutions based on silica optical fiber. Additionally, a team of researchers from Shanghai presented a sensing platform based on FBGs using the fused deposition modeling (FDM) method for the construction of the structure [53]. This platform was composed of several cylindrical structures in polylactic acid (PLA) with the FBG inside them. This device was designed to be used as a fixed broad platform for plantar pressure monitoring, which demonstrated to have a reliable mechanical performance.

Finally, a noninvasive and efficient insole FBG-based architecture for monitoring plantar pressure was presented in [32]. This work stands out from the others, because the authors introduced a whole IoT solution with the insole sensors integrated with a wireless transceiver, exhibited high energy efficiency and secured data transmission, to ensure the mobility and privacy of user data. The presented data reflected the precision of the proposed system, with the sensors having sensitivities up to 7.8 pm/kPa.

3.2 Plantar pressure and shear sensors

FBGs have also great potential for measuring shear stresses in the shoe. Although there are reports of sensors developed for shear measurement [54, 55], measuring plantar and shear pressure simultaneously is more attractive and provides more insights about the wellbeing of the foot and the overall health of the person. Due to the advantages of fiber optics, and FBGs in particular, several researchers have been working on the design of FBG-based cells able to measure these two forces simultaneously. Although the main objective is the measurement of these parameters during gait, none of the studies refers the introduction of the developed sensing cells in insoles or platforms, presenting only the cells in its isolated form.

The first work published with simultaneous shear and vertical forces sensing with FBGs goes back to the year of 2000. The team of Koulaxouzidis developed a cell, using three optical fibers with an FBG each, embedded in a block of elastic material, as schemed in **Figure 8a** [56]. The developed sensor was able to measure the vertical stress, as well as the magnitude and direction of the shear stress on its top surface. The experimental results showed a good repeatability and a resolution near to 5 kPa in the measurement of both forces. Later, in 2013, Zhang et al. developed an identical sensor to the previous one. In this case, two POFs with one FBG each were used, one of them was placed horizontally (hPOF, hFBG), while the other was tilted (tPOF, tFBG). Both fibers were embedded in a soft polydimethylsiloxane (PDMS) matrix, as shown in **Figure 8b** [57]. The sensor had a 27 mm length and width, and a 22 mm height. In this work, the obtained pressure sensitivity was 0.8 pm/Pa in a full range of 2.4 kPa, and the shear stress sensitivity was 1.3 pm/Pa for a full range of 0.6 kPa.

In 2015, Chethana et al. developed an optical sensor ground reaction force measurement platform for gait and geriatrics studies [58]. The developed system consisted of eight FBGs to measure the respective soil reaction forces on the three axes (x, y, and z). Four of the FBGs were placed at the vertices of the measuring platform, monitoring the shear motions on the x and y-axes (two for the x-axis and two for the y-axis motions detection). The remaining four FBGs were placed one on each frame supporting leg to measure the plantar pressure exerted on these zones. According to the authors, the optical fiber sensors platform for ground reaction force measurements presented a zero cross-force sensitiveness in all three loading axes [58].

In 2018, Tavares et al. developed a cell with the same operating principle as previously reported, but using only one silica optical fiber with two FBGs placed individually in two adjacent cavities, one made of cork and another of polylactide acid (PLA), as shown in **Figure 9** [59]. For the cells' calibration, the used method was similar to the one described in Ref. [57], and the obtained values were

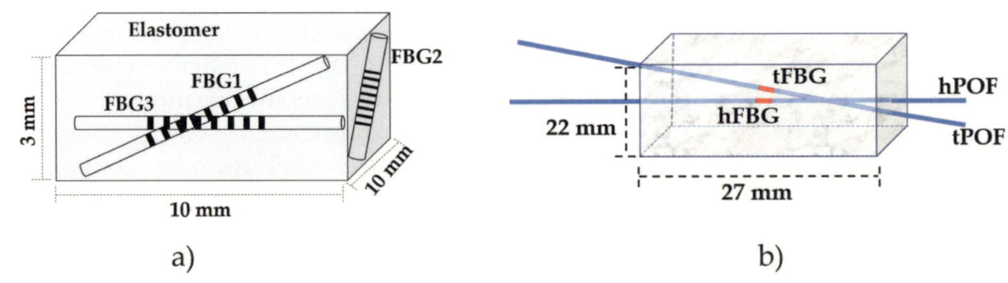

Figure 8.
Schematic representation of the FBG-based sensor cell developed in (a) silica (adapted from [56]) and (b) POF (adapted from [57]) fibers.

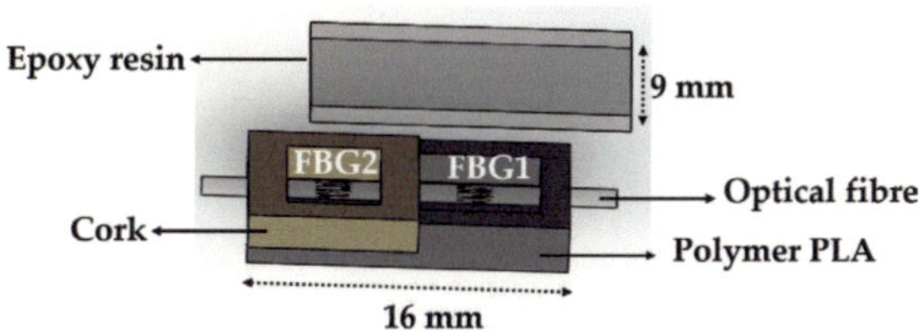

Figure 9.
Scheme of the shear and pressure sensing cell with its different components and respective dimensions (adapted from [59]).

compared with a 3-axial electronic force sensor. The results demonstrated that the developed device is a reliable solution for simultaneous measurement of shear and vertical forces. This solution has a great advantage over previous ones, since it only requires one optical fiber, which facilitates its incorporation into insoles. Therefore, several points, along the foot plantar surface, can be measured with a single optical fiber [12, 51, 52], with the advantage of being able to simultaneously diferentiate the two diferent forces (shear and pressure).

There are also studies using FBGs for vertical and shear forces measurements, but in which shear measurement is indirectly inferred from temperature variations. The authors argue that a rise in temperature in a certain area of the foot presupposes that there was friction between the surface of the foot and the shoe (shear force) [44, 60]. Najafi's team published a work in 2017 with the validation of a smart-textile based on fiber-optics with FBGs (SmartSox) for simultaneous measurement of temperature, pressure, and joint angles in patients with diabetic peripheral neuropathy (PND), where irregular temperature increase suggested the presence of shear forces [44]. In this study, FBG sensors were placed in socks that were successfully tested in a clinical setting by 33 individuals with PND to evaluate plantar pressure and temperature during normal gait velocity in a clinical setting.

4. Lower limb joints monitoring

The knee, hip, and ankle have a key role in gait, as it allows the body locomotion with muscles' minimum energy consumption and provides stability to walk in different terrain relief. During gait, the lower limb joints act together in order to provide the smoothest locomotion for the body. In **Figure 10**, the kinematics of the lower limbs in the different phases of gait are represented, namely the stance (a) and the swing phases (b) [14, 28].

At the beginning of the stance phase, in first double support and at the heel strike, the hip is flexed at 30°, the knee is extended and the ankle is at a neutral position. As the loading response approaches with the foot flat, the hip continues in a flexed mode as the knee starts flexing 5–10°, along with the ankle plantar flexing up to 20°, for the weight acceptance, shock absorption, and to propel the body forward. At the mid stance, the hip is extended, the knee flexed by 5–10° and the ankle is dorsi-flexed, with the purpose to move the body over the stationary foot. As the heel rises, with the ankle dorsi-flexing at 15°, the hip is extended at 15–30° and the knee is extended and then flexing. At the last point of the stance phase, in the toe off moment, the hip is flexing, the knee is also flexing at 30–40° and the ankle has a plantar flexing of 20–30°, in the preparation for the swing phase and the transfer of the load to the other limb [61].

At the initial swing, the hip continues flexing at 15°, the knee is flexing up to 65°, and the ankle is plantar flexed at 10° to clear the foot from the floor and advance the

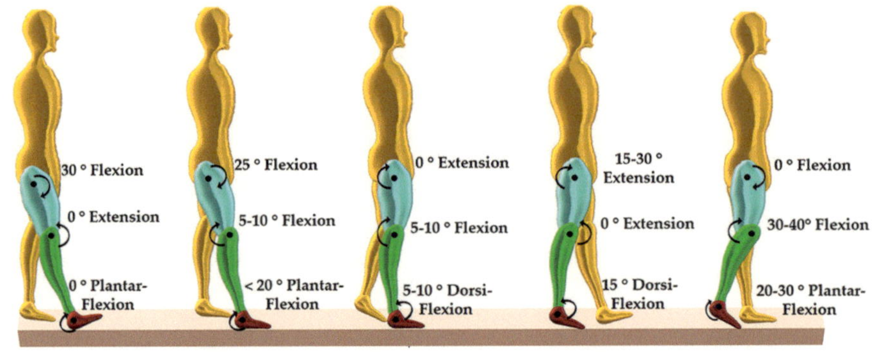

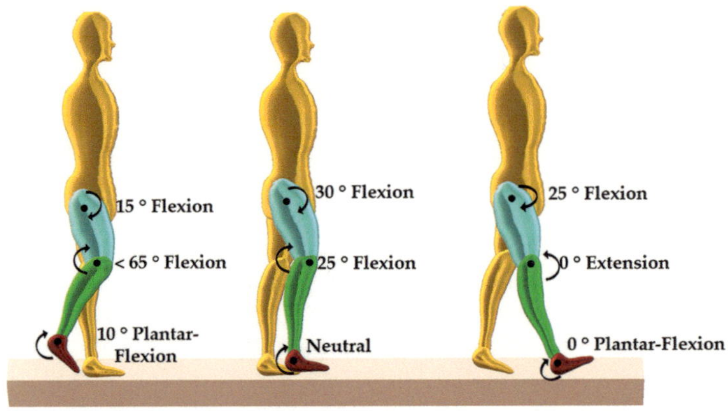

Figure 10.
Schematic representation of the lower limbs kinematics involved in the: (a) stance and (b) swing phases.

limb. At the mid swing, the hip is flexed at 30°, while the knee is flexing at 25°, and the ankle is at a neutral position. At the terminal swing, the hip is flexed at 25°, the knee is extended, and the ankle is at a 0° plantar flexion, to prepare the next heel contact at the beginning of the new stance phase [61].

The use of objective techniques to evaluate the health conditions of the knee can be a powerful tool for researchers and medical staff, providing relevant information about tendon-ligament strains and vibration, pressure, angular range of movements, and even temperature [62, 63].

There are numerous conventional techniques that can be used to monitor the joint conditions, such as stereo-optic, solid state, and piezo-resistive sensing methods, which employ accelerometers, magneto-resistive sensors, flexible goniometers, electromagnetic tracking systems, among others [24, 62, 64]. However, these techniques usually require complex and expensive electronics, which are susceptible to magnetic interferences and also cannot be used in humid/wet environments. Therefore, they do not represent an ideal solution for wearable sensing configurations, where the human transpiration may influence the sensors performance. So, the increasing research in the field of optical fiber sensors has also been focusing in the introduction of FBG technology in monitoring the lower limb joints during walking. Optical fiber sensors can be easily adapted to curved surfaces and various contours of the human body, especially the knee, a joint with complex anatomy [62, 63, 65–67].

The ideal technology to monitor limb joints should be able to measure curvature, being useful not only to monitor the motion of the lower limb segments, as well as to evaluate all the corporal posture [65]. The development of a smart garment, based on FBGs and flex sensing technologies, to monitor the body posture and lower and upper limbs' movements, was reported. An FBG-based sensing belt was produced by encapsulating FBG sensors inside a synthetic silica gel, as depicted in **Figure 11**, which was afterward attached to a garment for monitoring joints and body posture. The encapsulation of the FBG was made with an applied pre-stress, so the sensor is able to monitor both extension and compression deformations. The proposed FBG belt, fixed near the limb joints, is influenced by the body posture shifts, and the consequent sensor's Bragg wavelength shift was correlated with the angles at the limb joint [65]. Although the results presented by Abro et al. are only related to tests made at the upper limbs, the reasonable results obtained within the tests and exercises are a good indication of its potential application for the monitoring of the lower limbs motion.

4.1 Knee flexion-extension monitoring

From the lower limb three joints (ankle, knee, and hip), the knee is one of the body joints most prone to develop osteoarthritis [68]. Therefore, the supervision and monitoring of the motion of the knee are of crucial importance in the medical and physical rehabilitation field [37, 62, 67, 69, 70].

Rocha et al. suggest a wearable knee motion sensor, designed with a single FBG, embedded in a stretchable band of polyvinyl chloride (PVC) material and placed in the center of the knee joint, as schematized in **Figure 12a** [67]. The objective is to measure the knee movements from the straight leg to the maximum knee flexion and to obtain graphically the pattern of human gait, by monitoring flexion and extension, with the joint acting as a rotation axis, as represented in **Figure 12b**. The PVC band with the FBG was attached to an elastic ribbon (knee brace), by metallic pressure-buttons that ensure the stability of the sensing band while walking or running. In the reported work, the authors tested the proposed solution on a treadmill, under different types of run and speed, accompanied by video recorder [67]. The video was used to define the starting time of the stance and swing phases in order to correlate the data provided by the FBG sensor to the different phases of the walking routine [67, 71, 72].

When the leg is straight, the FBG sensor is in the resting position since there is neither flexion nor extension in the optical fiber. Once the bending movement of the knee starts, during walking, it results in an extension of the optical fiber, inducing a strain in the FBG sensor, positioned at the center of the knee joint. Consequently, a positive shift of the reflected Bragg wavelength is obtained. The reverse bending movement, from the maximum knee flexion point to straight leg, leading to a relaxation of the FBG, return to its initial Bragg wavelength value [67]. By monitoring the wavelength shift during these movements, the gait pattern of the patient could be characterized.

Although the researchers Rocha et al. show a clear characterization of the movement of the knee joint during the gait cycle, they also point out, as a drawback, the noise induced in the signal by vibration, considering that better results are achieved at lower speed, softening the influence of the elastic factor of the knee band [67].

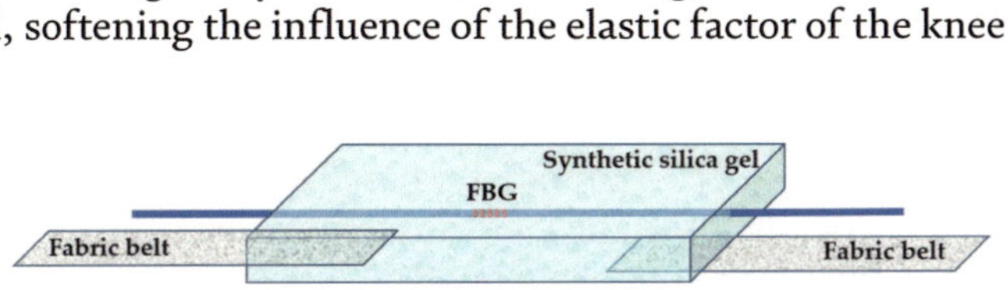

Figure 11.
Schematic representation of the FBG belt proposed by Abro et al. (adapted from [65]).

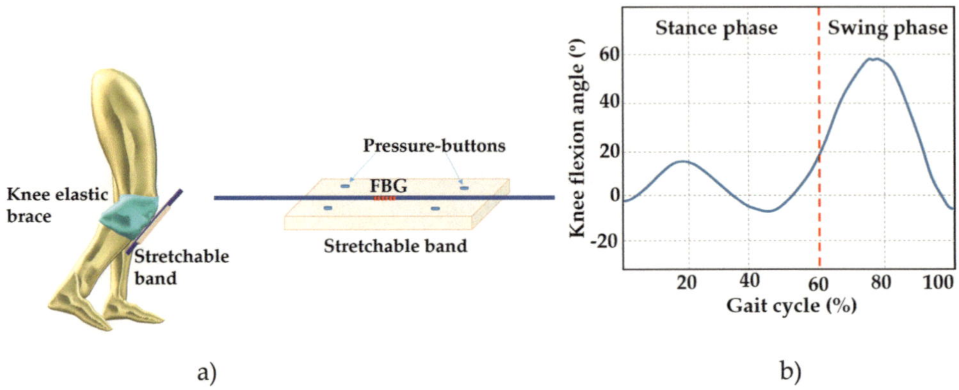

Figure 12.
(a) Schematic representation of an FBG-based solution for knee movements monitoring (adapted from [67]); and (b) typical keen angle pattern during gait.

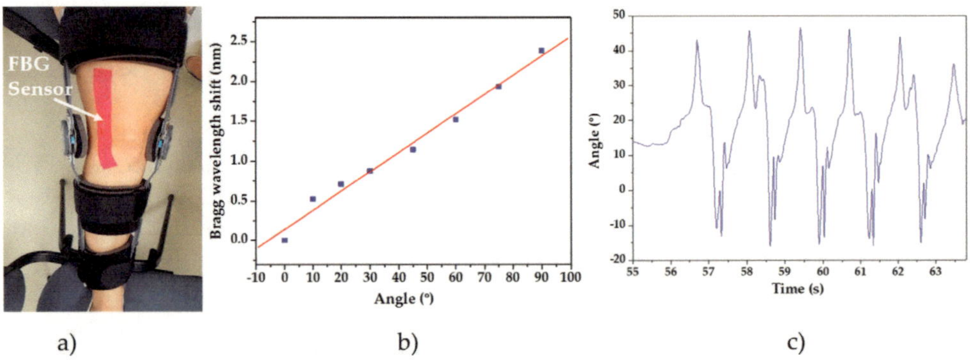

Figure 13.
(a) Photograph of the kinesio tape with an embedded FBG for keen angle monitoring; (b) Bragg wavelength shift dependence with the keen flexion angle; and (c) knee flexion/extension angle during six gait cycles.

Similar results can be achieved using kinetic tape (elastic adhesive tape) with an embedded FBG. The kinetic tape is attached to the lower limb, starting at the quadriceps area and ending at the beginning of the tibia, with the FBG placed just a few centimeters above the knee rotation axis, as shown in **Figure 13a**. Such configuration is a more stable solution, since the fiber containing the FBG is only actuated by the rotation of the knee, which stretches the kinetic tape inducing a strain and consequent positive wavelength shift in the FBG. During the calibration process, using an angle lock goniometer for angles ranging between 0 and 90°, a direct relation between the knee angle and the Bragg wavelength shift was found as displayed in **Figure 13b**. In **Figure 13c** is presented the flexion/extension angles, along time for six gait cycles, obtained with the solution represented in **Figure 13a**, and which as a similar behavior as reported by the authors in Ref. [67], but with a considerable reduced noise level.

4.2 Ankle flexion and dorsi-flexion monitoring

Umesh et al. proposed an FBG goniometer based on the deflection produced in an optical fiber by variation of the angle of the goniometer [73]. The purpose of the sensor is to measure the range of movement (ROM), which for the ankle joint the movement can be classified as ROM plantar-flexion and ROM dorsi-flexion. Plantar flexion is described as the rotation that increments the angle described between foot and the shin, and the dorsi-flexion is the rotation that results in a lower angle. The sensor is an assembly of two discs of 30 mm, overlapped by two discs with smaller diameter (5 mm). The two pairs of discs are circled by a rubber belt, to ensure synchronized rotation between them. The optical fiber with the FBG sensor is placed in a cantilever, connected to the upper belt. The rotation arm is linked to the side of the foot and its movement motivates the rotation of the correspondent disc. This rotation moves the cantilever and creates strain in the FBG, which can be rewritten in angle values, by proper calibration. The characterization of these two rotations has crucial importance in clinical diagnosis, helping the evaluation of the limitations of this joint. Furthermore, it is a noninvasive method of measurement with the advantages that optical fibers offer, and that can counteract to limitations of conventional electro-goniometers and video tracking systems as electromagnetic interference, size, and fragility [73].

4.3 Tendons and ligaments monitoring

Beyond their ability to measure the flexion, dorsiflexion, and extension of the joints involved in gait, FBGs can broaden their usage to applications related to the tendons and muscles monitoring. Although it may fall a bit out of the scope of e-Health, it is worth mentioning the application of FBGs to perform pressure mapping, and monitor strain and length of tendons and ligaments, when under load or locomotion. Ren et al. presented an FBG sensor embedded in a micro-shape memory alloy tube which is able to measure the displacement of the tendon [74]. To verify the performance of the sensor, the initial tests were made in the Achilles tendon and the results compared with the ones obtained simultaneously with a two-camera stereovision sensor. The fiber-based sensor was also applied to a cadaver knee tendon, in the medial and lateral collateral ligament, to record the deformation of the ligaments in simulated postures. The results proved that the FBG sensor has high sensitivity and low signal-to-noise ratio, without loss of accuracy. It is also easily implemented and minimally invasive to the biological tissues, projected to be applied *in-vivo*, after some improvements [74].

5. Prosthetic and exoskeletons applications

For severe impaired citizens, it is common to adapt prosthetic lower limbs, in order to offer patients mobility. The interface pressure inside the prosthetic sockets is of major relevance, in order to avoid ulcerations in the patients and evaluate its suitability. Moreover, the application of robotics technology to improve the wellbeing of debilitated patients has been highly investigated in the past few years. In particular, exoskeletons can be wearable devices prone to be used to restore functional movements of amputees and persons with paralysis. Therefore, this section surveys the use of FBG sensors for the development and evaluation of prosthetic limbs, in addition to control and automation of exoskeletons.

5.1 Prosthetic limbs

The partial or total limb amputation is one of the oldest treatment options available in medicine. Unfortunately, the frequency of the lower limb amputation is growing worldwide. Traffic accidents, particularly motorcycle accidents, health problems (including diabetes, arterial hypertension, chronic renal failure, and hypercoagulability), and advanced age are the main causes. Additionally, this is a predominant incident in countries affected by landmines and other natural disasters, including, for instance, earthquakes. Due to the socioeconomic impact (with the consequent inability to work and socialize), the interference on the life quality, and other complications, such as hematoma, infections, necrosis, contractures, neuromas, and phantom pain; this is a relevant public health problem.

The key element of amputee rehabilitation is the engineering of solutions, appropriated for individuals to recover their physical capabilities. A prosthesis or artificial limb is a device, whose function is to substitute the limb that was lost, with cosmetic and functionality for the amputee. A lower limb prosthesis results from the assembly of several components, including socket, shank, ankle, and foot, as schematized in **Figure 14**.

The socket is the most relevant component of the artificial limb, since it constitutes the critical interface between the amputee's stump and the amputee. The design and fitting of the socket are also the most difficult processes, due to the particularity of each amputee's stump. When wearing the prostheses, the appropriated fit and comfort are critical factors that contribute to its successful use. Nevertheless, many amputees still complain about discomfort or pain, reporting a set of problems, including edema, pressure ulcers, dermatitis, and skin irritation, due to the use of the prostheses [76]. This is particularly related with the changes in the residual limb soft tissues (volume, shape, sensitivity, composition, among others), which vary during the day due to factors such as temperature, activity, and hydration.

As result, in the last years, several measurement systems have been proposed to assess the interface pressure between the residual limb and the prosthetic socket [77]. This includes electrical strain gauge [78], F-socket transducer arrays [79, 80], and finite element analysis [81–83]. The output from these systems has been used to improve the socket design. Nevertheless, despite the technological advances in the existing socket design and the measurement systems, available sockets still exhibit many weaknesses. For instance, apart from the high accuracy and sensitivity provided, the use of strain gauges requires modifying the sockets with openings for accommodation of the device. This procedure interferes in the socket shape, and consequently in the accuracy of the pressure measurements.

In the case of the F-sockets, these systems present flexibility, good sensitivity, and ease of use. Additionally, in contrast to strain gauges, no change in the

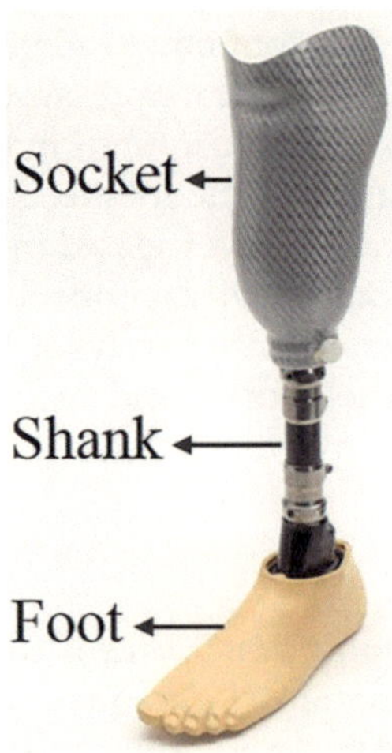

Figure 14.
Typical transtibial prosthesis (adapted from [75]).

socket shape is required, since F-sockets are quite thin, which can be placed *in-situ* between the residual limb and the prosthetic socket. Nevertheless, the nonlinearity, hysteresis, drift, and vulnerability to electromagnetic interferences are the main limitations. Additionally, the shear stresses are not accounted for, when this system is used.

The finite element analysis is a numerical modeling alternative, which, when applied to the residual limb, predicts the soft tissues load distributions and magnitudes. This information has been assisting the technicians during the socket design. Nonetheless, although some models already considered thresholds for tissue injury and adverse adaptation, and other researchers have included in the models parameters, such as comfort and pain threshold, several complaints are still reported from the use of the prostheses, due to the subjectivity, difficulty to evaluate these factors, and the inter- and intra-individual loading [83].

Consequently, new sensing methodologies with minimal limitations toward accurate measurements of the interface pressure within prosthetic sockets are essentially required. Thereby, the FBG technology was pointed out as a potential alternative to conventional methodologies [84]. In 2010, Kanellos et al. proposed a 2D optical FBG-based pressure sensor, predicting to be suitable for several biomedical applications, namely biomechanics, rehabilitation, and orthotics, including amputee sockets [85]. The device consists of FBGs embedded into a thin polymer layer of PDMS, with the minimum thickness of the sensing pad set to 2.5 mm. The sensor exhibited a maximum fractional pressure sensitivity of 12 MPa^{-1}, with a spatial resolution of $1\times 1\ cm^2$, also revealing no hysteresis and real-time operation possibility. Due to the elasticity and ductility of the polymer, which match human skin behavior, the system becomes a flexible 2D pressure sensing surface. This configuration is appropriate to be attached or anchored to irregular shaped objects/bodies, allowing to translate more accurately all the phenomena that may occur in them. These properties meet the requirements of human machine interfaces, comprising amputee sockets, as initially predicted.

Apart from the medical requirements imposed on the FBG embedded-based sensor pads, which include real time acquisition, high sensitivity and resolution, and increased dynamic range, these systems also need to comply with a set of demands related to fabrication and packaging processes. These conditions result from the diversity of the patients to be treated, and also from their real life conditions. Thus, the influence of the fiber embedding depth (center and top position of pad cross-section), the thickness of the polymer sheet (2 and 3 mm), and the fiber type (hydrogenated SMF-28 and nonhydrogenated GF1B) were assessed in [86]. The results of this study reveal that the sensor pads rigidity and durability are enhanced, when the Bragg grating, inscribed into nonhydrogenated fiber, is embedded at the polymer center, with a thickness of 3 mm.

Results of the first investigation of the ability of FBGs to measure interface pressure between the stump of a trans-tibial amputee and a patellar tendon bearing (PTB) prosthetic sockets are presented in [75]. The patellar tendon (PT) bar was the key analyses' area since this supports the majority of the transtibial amputee's body weight, when the subject is using the PTB socket. In Ref. [75], the FBGs were first embedded into an epoxy material (NOA 61), aiming to acquire the required protection to withstand the high pressure values up to 230 kPa at the PT bar [78]. After that, this sensing pad was placed between two silicone polymeric sheets forming the pressure sensor, as schematized in **Figure 15**.

Since the initial contact of the PT with the sensor surface is mostly pressure concentrated, this behavior was imitated using a ball bearing, and positive wavelength shift of 3.8 nm was observed for a maximum load of 30 N, **Figure 16a**.

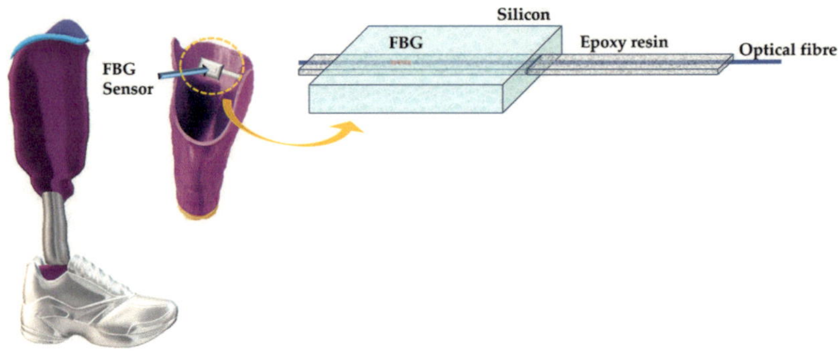

Figure 15.
Schematic representation of an FBG-based system for monitor the interface pressure between the socket and the amputee (adapted from [75]).

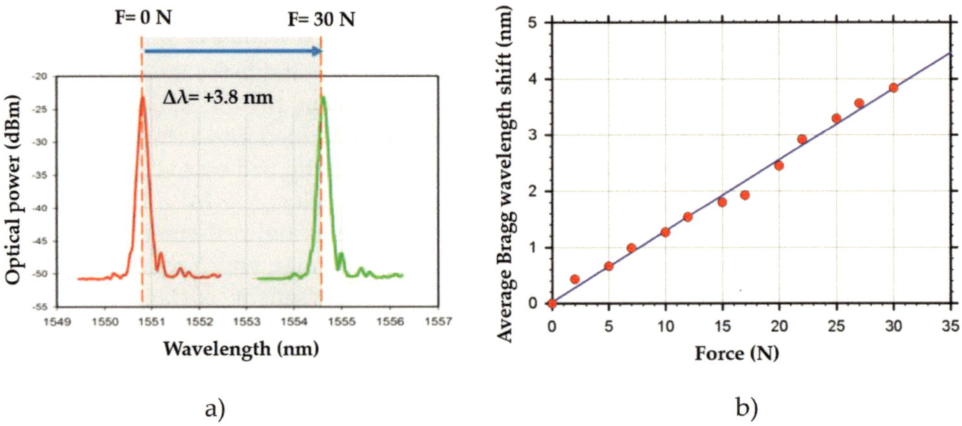

Figure 16.
(a) Representation of the maximum Bragg wavelength shift for maximum load applied (adapted from [75]); and (b) average Bragg wavelength shift as function of the applied force (adapted from [75]).

Thereafter, an experimental set up was designed to assess, *in-situ*, the sensor performance, while attached to the inner socket wall. Although there was no subject involved in these tests, consideration was taken to reproduce a real-life situation, as much as possible. The results obtained for the different load cycles reveal the suitability of the sensor to accomplish pressure measurements on the socket stump interface, especially in the PT region. From the calibration procedure, a proportional wavelength shift dependence with the load applied was found **Figure 16b**.

Toward using these sensors *in-situ*, the performance of these sensing pads was broadly assessed concerning the sensitivity, durability, and hysteresis error [87]. Similar to the work of Kanellos et al., three production parameters were investigated, which are the FBG embedding depth (top, bottom, and neutral layers of the sensing pad), the sensing pad thickness (1, 2, and 3 mm), and the type/hardness of sensing pad materials [PDMS (harder) and silicone rubber (softer)]. The best sensor's performance (highest sensitivity and accuracy) was obtained for the FBG embedded in the neutral layer of PDMS and with the thicker sensing pads. An FBG array was produced with these conditions and used for interface pressure measurements within prosthetic sockets. Additionally, to further assess the performance of the proposed sensing pad, these were evaluated *in-situ*, in a traumatic transtibial amputee using a total surface bearing socket, with 6 mm silicone liner. The results were validated comparing the data obtained with the FBG technology to the pressure measurements acquired by the F-socket sensors. Although the data obtained for the 8 sub-regions of the amputees' residual limb follow the same tendency; higher pressure values were registered by the FBG sensors. The difference was attributed to the sensors' thickness, which is 3 mm in the case of the FBG sensing pads and 0.2 mm in the F-socket sensing mats.

Aiming to eliminate the previous limitations and provide a simpler and more practical sensing procedure, Al-Fakih et al. proposed an innovative customized FBG-instrumented silicone liner, which consists of two silicone layers with 12 FBGs embedded between them, with the gratings located in clinical interest points [88]. In this study, a custom gait simulation machine was built to test the performance of the sensing system during an amputee's simulation gait. The data were validated with the findings obtained using an F-socket. The FBG technology revealed sensitivity and accuracy similar to the ones obtained with the F-socket technology. Nevertheless, this new design can be used repeatedly in clinical and research setting, which is an important benefit compared to the F-socket mats that, due to drift and calibration issues, are usually discarded after each utilization.

Recently, the technologies used to assess the interface pressure between the residual limb and the prosthetic socket, and the challenges found concerning the development of new solutions of sockets for limb prostheses were reviewed in [89]. In this chapter, FBGs are pointed out as one of these technologies. Additionally, the study observes that due to the high risk of the damaging of fibers, their applications are still limited, with further studies still required to confirm their suitability in this field. Nevertheless, the shown advantages of this technology over other sensing methodologies, especially regarding drift and linearity, and the constant low satisfaction level of the amputees, are enough motivations to continue investing on this technology.

5.2 Rehabilitation exoskeletons

The application of robotics, in particular robotic exoskeleton systems, to improve the wellbeing of debilitated patients is already being adopted. This technology is being used in human power augmentation, and its application has become more prominent, as to provide alternative solutions for physically limited people support in their daily movements [90].

Exoskeletons are known to be wearable robots (robotic exoskeletal structures), with a strict physical and cognitive interaction with the human user, since, typically, it operates alongside human limbs. Although the scientific and technological research on the development and implementation of exoskeletons began in the early 60s, only recently, its application in rehabilitation and functional substitution of movements have been implemented in patients with motor disorders [91].

Robotic exoskeletons provide unique methods for rehabilitation, by promoting the patient engagement in its training, and retrieving better quantitative feedback and improved functional outcome for patients. In a future perspective, the development of more effective exoskeletons is insight, with solutions for a real-time biological synergy-based exoskeleton, which will allow disabled patients to regain normal mobility capabilities [92, 93].

The exoskeleton feedback is based on the information, which is retrieved from the embedded sensors in its structure. The current exoskeleton designs can have up to several dozens of sensors, to monitor variables, such as rotation, torque, tilt, pressure, position, velocity, neurological signals, among others. As the sensing systems integrated in the robot are the key devices for its proper performance, the research field on robotics already has a mature and overspread technology, offering good sensitivity, precise measurements, and competitive price, with sensing systems often based on solid-state sensing [64]. Nevertheless, this technology has also shown some drawbacks, due to its susceptibility to electromagnetic interferences resulting from the electric inertial motors. This interference could be enough to degrade the sensors signal, sending erroneous information to the control devices, leading the exoskeleton to perform erratically, and risking injuring the patient.

Rehabilitation robotics applications also require the analysis of the body motion, in order to close control loops around defined joints. Commercial optical systems, such as Vicon, are considered the standard in human motion analysis. Although Vicon provides accurate position information, it has some significant limitations, such as high costs and limited measure volume, since it has to be used in laboratories with fixed equipment, which prevents its use in rehabilitation robotics applications [94]. On the other hand, soft exoskeletons require even more imperceptible sensors, typically sensor heads with thicknesses below 0.5 mm, in which electronic devices present some drawbacks, including long term instability, inconsistency, excessive drift, and the restriction to a small sensing area requiring the use of more sensors to monitor larger areas [64]. As an alternative to these electronic and optical sensors, the optical fiber sensors offer a small and robust solution, able to acquire kinematic and kinetic measurements, enhancing the exoskeleton performance by adding further responsiveness, controllability, and flexible motion. Nevertheless, the use of FBG sensors in exoskeletons is not yet widely explored, with only a limited number of studies reported. Recently, Domingues et al. reported the instrumentation of an insole with FBG sensors for plantar pressure monitoring [12, 32]. The reported wearable device is able to be adapted to exoskeletons structures, and dynamically retrieve the gait pattern of the patient.

Although there is a shortage of studies regarding the adaptation of FBG sensing technologies to exoskeletons, for gait aid there are already some reports focusing on its application in robot fingers and glove-based devices [95–97]. Park et al. presented an FBG-based solution to monitor the force in exoskeleton fingers [95]. The authors embedded the optical fiber sensors in a finger-like plastic 3D mesh, inspired in the design of arthropod limbs, near the fingers base, for enhanced sensitivity. With the developed structure, it is possible to detect forces down to 0.02 N, with a resolution of ~0.15 N. The robot hand instrumented by Park et al. was able to be operated in a hybrid control scheme, with the fingers being capable to sense small forces, with the advantage of being able to have all the FBG sensors in one single fiber, due to FBGs multiplexing ability [95].

Jiang et al. also described the design and production of an instrumented robotic hand with three fingers that enable both pinch and power grips. The optical FBG sensors were embedded in both the rigid plastic and soft skin material that constitutes the hand bone structure. In the rigid plastic material, the authors included eight FBGs for force sensing, while in the soft skin, they integrated six FBGs strain sensors for tactile monitoring, providing information on the location of the contact points [96]. Although there are already some studies related to the upper limbs motion aid, some work is still needed concerning the application of FBG technology to exoskeletons for gait rehabilitation of patients, which demands a direct focus on the lower limbs synergy between the patient and the exoskeleton.

Key topics for further development of exoskeletons in rehabilitation scenarios include the need for robust human-robot multimodal cognitive interaction, safe and dependable physical interaction, true wearability and portability, and user aspects such as acceptance and usability [91]. It should be able to augment the ability and/or to treat skeletal parts, which are weak, ineffective, or injured due to a disease or a neurological condition. Therefore, the exoskeleton should be designed to work in parallel with human body and be actuated either passively and/or actively [98].

6. Conclusion

e-Health has been widely investigated in recent years, building on technological advances, especially in fields such as sensing and networking. Building on such gains, more innovations are expected to enhance the life quality of citizens, especially debilitated and elder ones. Gait analysis stands out as one promising solution, which can help in the rehabilitation of locomotive impairments, in addition to early diagnosis of other pathologies, such as ulcers in patients with diabetes. Various solutions have been proposed in the literature for close monitoring and analysis of gait. However, recently, FBGs have been pointed out as a promising alternative for a sensing technology to analyze gait movement, building on advantages such as small size, rigidness, low-cost, low power consumption, and minimally invasive. Due to its recent adoption and promising advantages, this chapter has provided a thorough review of research and design efforts of FBG-based sensors for gait analysis. The chapter initially explains the sensing principle underlying the FBG technology, after that the topic of gait analysis and the different phases of gait cycle are described, and then moves toward required e-Health monitoring solutions. Efforts toward the design of solutions to monitor plantar pressure and shear forces are discussed. Monitoring of plantar pressure, independently, is first presented, then simultaneous monitoring of plantar and shear forces is further elaborated. The chapter then moves toward monitoring of lower limb joints, which also play key roles in the gait analysis, since their wellbeing affects the gait cycle pattern. The use of optical fiber sensing in prosthetic and exoskeletons concludes the topics discussed in the chapter. This chapter represents a thorough review of research efforts in the design of optical fiber-based sensors in gait analysis, covering all related topics of monitoring plantar pressure, shear forces, knee and joints, and integration in prosthetic and exoskeletons.

Acknowledgements

This work is funded by FCT/MEC through national funds and when applicable co-funded by FEDER – PT2020 partnership agreement under the projects, UID/EEA/50008/2013, UID/CTM/50025/2013 and 5G-AHEAD IF/FCT- IF/01393/2015/CP1310/CT0002. Nélia Alberto acknowledges PREDICT (FCT-IT-LA) scientific action; Cátia Tavares acknowledges her PhD grant PD/BD/142787/2018. The financial support from FCT through the fellowships SFRH/BPD/101372/2014 (M. Fátima Domingues) and SFRH/BPD/109458/2015 (Carlos Marques) is also acknowledged.

Author details

Maria de Fátima Domingues[1*], Cátia Tavares[2], Tiago Leite[1,2], Nélia Alberto[1], Cátia Leitão[1,2], Carlos Marques[1,2], Ayman Radwan[1], Eduardo Rocon[3], Paulo Antunes[1,2] and Paulo André[4]

1 Instituto de Telecomunicações, Campus Universitário de Santiago, Aveiro, Portugal

2 Department of Physics & I3N, University of Aveiro, Campus Universitário de Santiago, Aveiro, Portugal

3 Centro de Automática y Robótica, CSIC-UPM, Madrid, Spain

4 Department of Electrical and Computer Engineering, Instituto de Telecomunicações, Instituto Superior Técnico, University of Lisbon, Lisbon, Portugal

*Address all correspondence to: fatima.domingues@ua.pt

References

[1] Kawasaki B, Hill K, Johnson D, et al. Narrow-band Bragg reflectors in optical fibers. Optics Letters. 1978;**3**:66-68. DOI: 10.1364/OL.3.000066

[2] Lam D, Garside B. Characterization of single-mode optical fiber filters. Applied Optics. 1981;**20**:440-445. DOI: 10.1364/AO.20.000440

[3] Hill K, Meltz G. Fiber Bragg grating technology fundamentals and overview. Journal of Lightwave Technology. 1997;**15**:1263-1276. DOI: 10.1109/50.618320

[4] Kashyap R, editor. Fiber Bragg Gratings. 2nd ed. San Diego: Academic Press; 2009. 632 p. ISBN: 9780080919911

[5] Meltz G, Morey W, Glenn W. Formation of Bragg gratings in optical fibers by a transverse holographic method. Optics Letters. 1989;**14**:823-825. DOI: 10.1364/OL.14.000823

[6] Hill K, Malo B, Bilodeau F, et al. Bragg gratings fabricated in monomode photosensitive optical fiber by UV exposure through a phase mask. Applied Physics Letters. 1993;**62**:1035-1037. DOI: 10.1063/1.108786

[7] Muro-de-la-Herran A, Garcia-Zapirain B, Mendez-Zorrilla A. Gait analysis methods: An overview of wearable and non-wearable systems, highlighting clinical applications. Sensors. 2014;**14**:3362-3394. DOI: 10.3390/s140203362

[8] Gehlsen G, Beekman K, Assmann N, et al. Gait characteristics in multiple sclerosis: Progressive changes and effects of exercise on parameters. Archives of Physical Medicine and Rehabilitation. 1986;**67**:536-539. DOI: 10.5555/uri:pii:0003999386905496

[9] Kerrigan D, Tidd M, Croce U, et al. Biomechanical gait alterations independent of speed in the healthy elderly: Evidence for specific limiting impairments. Archives of Physical Medicine and Rehabilitation. 1998;**79**:317-322. DOI: 10.1016/S0003-9993(98)90013-2

[10] Stolze H, Klebe S, Getersen G, et al. Typical features of cerebellar ataxic gait. Journal of Neurology, Neurosurgery and Psychiatry. 2002;**73**:310-312. DOI: 10.1136/jnnp.73.3.310

[11] Afilalo J, Eisenberg M, Morin J-F, et al. Gait speed as an incremental predictor of mortality and major morbidity in elderly patients undergoing cardiac surgery. Journal of the American College of Cardiology. 2010;**56**:1668-1676. DOI: 10.1016/j.jacc.2010.06.039

[12] Domingues M, Tavares C, Leitão C, et al. Insole optical fibre Bragg grating sensors network for dynamic vertical force monitoring. Journal of Biomedical Optics. 2017;**22**:091507(8pp). DOI: 10.1117/1.JBO.22.9.091507

[13] Sutherland D. The evolution of clinical gait analysis part I: Kinesiological EMG. Gait and Posture. 2001;**14**:61-70. DOI: 10.1016/S0966-6362(01)00100-X

[14] Sutherland D. The evolution of clinical gait analysis. Part II kinematics. Gait Posture. 2002;**16**:159-179. DOI: 10.1016/S0966-6362(02)00004-8

[15] Sutherland D. The evolution of clinical gait analysis part III–Kinetics and energy assessment. Gait and Posture. 2005;**21**:447-461. DOI: 10.1016/j.gaitpost.2004.07.008

[16] Mummolo C, Mangialardi L, Kim J. Quantifying dynamic characteristics of human walking for comprehensive gait cycle. Journal of Biomechanical

Engineering. 2013;**135**: 091006 (10pp). DOI: 10.1115/1.4024755

[17] Soames R. Foot pressure patterns during gait. Medical Engineering and Physics. 1985;7:120-126. DOI: 10.1016/0141-5425(85)90040-8

[18] Marasović T, Cecic M, Zanchi V. Analysis and interpretation of ground reaction forces in normal gait. WSEAS Transactions on Systems. 2009;**8**: 1105-1114. ISSN: 1109-2777

[19] Perry J, Davids J. Gait analysis: Normal and pathological function. Journal of Pediatric Orthopaedics. 1992;**12**:815. DOI: 10.1097/01241398-199211000-00023

[20] Vaughan C, Davis B, O'Connor J, editors. Dynamics of Human Gait. 2nd ed. South Africa: Kiboho Publishers; 1999. 137p. ISBN: 0-620-23558-6

[21] Perttunen J. Foot loading in normal and pathological walking [thesis]. Jyväskylä: University of Jyväskylä; 2002

[22] Prakash C, Kumar R, Mittal N. Recent developments in human gait research: Parameters, approaches, applications, machine learning techniques, datasets and challenges. Artificial Intelligence Review. 2018;**49**: 1-40. DOI: 10.1007/s10462-016-9514-6

[23] Whittle M. Gait analysis. In: McLatchie G, Lennox C, editors. The Soft Tissues – Trauma and Sports Injuries. Oxford: Butterworth-Heinemann; 1993. pp. 187-199. ISBN: 0 7506 0170 1

[24] Tao W, Liu T, Zheng R, et al. Gait analysis using wearable sensors. Sensors. 2012;**12**:2255-2283. DOI: 10.3390/s120202255

[25] Shi DL, Wang YB, Ai ZS. Effect of anterior cruciate ligament reconstruction on biomechanical features of knee level in walking: A meta-analysis. Chinese Medical Journal. 2010;**123**:3137-3142. DOI: 10.3760/cma.j.issn.0366-6999.2010.21.034

[26] Langer S. A Practical Manual of Clinical Electrodynography. 2nd ed. New York, USA: The Langer Foundation for Biomechanics and Sports Medicine Research; 1989. ISBN: 0936445025

[27] Kyriazis V. Gait analysis techniques. Journal of Orthopaedics and Traumatology. 2001;**2**(1):1-6. DOI: 10.1007/PL00012205

[28] Terrier P, Schutz Y. How useful is satellite positioning system (GPS) to track gait parameters? A review. Journal of NeuroEngineering and Rehabilitation. 2005;**2**:28(11pp). DOI: 10.1186/1743-0003-2-28

[29] Frigo C, Rabuffetti M, Kerrigan D, et al. Functionally oriented and clinically feasible quantitative gait analysis method. Medical and Biological Engineering and Computing. 1998;**36**:179-185. DOI: 10.1007/BF02510740

[30] Al-Fakih E, Osman N, Adikan F. The use of fibre Bragg grating sensors in biomechanics and rehabilitation applications: The state-of-the-art and ongoing research topics. Sensors. 2012;**12**:12890-12926. DOI: 10.3390/s121012890

[31] Domingues M, Tavares C, Alberto N, et al. Non-invasive insole optical fibre sensor architecture for monitoring foot anomalies. In: Proceedings IEEE Global Communications Conference (GLOBECOM 2017); 4-8 December 2017; Singapore. New York: IEEE; 2018. 6p. DOI: 10.1109/GLOCOM.2017.8255026

[32] Domingues M, Alberto N, Leitão C, et al. Insole optical fibre sensor architecture for remote gait analysis-an eHealth solution. IEEE Internet of Things Journal. DOI: 10.1109/JIOT.2017.2723263

[33] Farahani B, Firouzi F, Chang V, et al. Towards fog-driven IoT eHealth: Promises and challenges of IoT in medicine and healthcare. Future Generation Computer Systems. 2018;**78**(part 2):659-676. DOI: 10.1016/j.future.2017.04.036

[34] Aging and Health, Fact sheet n°404 (World Health Organization) [Internet]. Available from: http://www.who.int/mediacentre/factsheets/fs404/en/ [Accessed: February 13, 2018]

[35] McWilliams A. Global markets and technologies for home medical equipment. bccResearch [Internet]. 2016. Available from: https://www.bccresearch.com/market-research/healthcare/home-medical-equipment-technologies-market-report-hlc054d.html [Accessed: July 27, 2018]

[36] Eskofier B, Lee S, Baron M, et al. An overview of smart shoes in the internet of health things: Gait and mobility assessment in health promotion and disease monitoring. Applied Sciences. 2017;**7**:986(17pp). DOI: 10.3390/app7100986

[37] Mignani A, Baldini F. Biomedical sensors using optical fibres. Reports and Progress in Physics. 1996;**59**:1-28. DOI: 10.1088/0034-4885/59/1/001

[38] Grattan K, Sun T. Fibre optic sensor technology: An overview. Sensors and Actuators: A. 2000;**82**:40-61. DOI: 10.1016/S0924-4247(99)00368-4

[39] Alberto N, Bilro L, Antunes P, et al. Optical fibre technology for eHealthcare. In: Cruz-Cunha M, Miranda I, Gonçalves P, editors. Handbook of Research on ICTs and Management Systems for Improving Efficiency in Healthcare and Social Care. Hershey, USA: IGI Global; 2013. pp. 180-200. DOI: 10.4018/978-1-5225-5484-4.ch069

[40] Demarré L, Lancker A, Hecke A, et al. The cost of prevention and treatment of pressure ulcers: A systematic review. International Journal of Nursing Studies. 2015;**52**:1754-1774. DOI: 10.1016/j.ijnurstu.2015.06.006

[41] Rahman S, Rahman T, Ismail A, et al. Diabetes-associated macrovasculopathy: Pathophysiology and pathogenesis. Diabetes, Obesity and Metabolism. 2007;**9**:767-780. DOI: 10.1111/j.1463-1326.2006.00655.x

[42] Cobb J, Claremont D. Transducers for foot pressure measurement: Survey of recent developments. Medical and Biological Engineering Computing. 1995;**33**:525-532. DOI: 10.1007/BF02522509

[43] Lord M, Hosein R. Study of in-shoe plantar shear in patients with diabetic neuropathy. Clinical Biomechanics. 2000;**15**:278-283. DOI: 10.1016/S0268-0033(99)00076-5

[44] Najafi B, Mohseri H, Grewal G, et al. An optical-fibre-based smart textile (smart socks) to manage biomechanical risk factors associated with diabetic foot amputation. Journal of Diabetes Science and Technology. 2017;**11**:668-677. DOI: 10.1177/1932296817709022

[45] Antunes P, Marques C, Varum H, et al. Biaxial optical accelerometer and high-angle inclinometer with temperature and cross-axis insensitivity. IEEE Sensors Journal. 2012;**12**:2399-2406. DOI: 10.1109/JSEN.2012.2190763

[46] Leitão C, Antunes P, Pinto J, et al. Carotid distension waves acquired with a fiber sensor as an alternative to tonometry for central arterial systolic pressure assessment in young subjects. Measurement. 2017;**95**:45-49. DOI: 10.1016/j.measurement.2016.09.035

[47] MacWilliams B, Armstrong P. Clinical applications of plantar pressure measurement in pediatric orthopedics.

In: Pediatric Gait: A new Millennium in Clinical Care and Motion Analysis Technology. 22 July 2000; Chicago, USA. New York: IEEE; 2002. pp. 143-150. DOI: 10.1109/PG.2000.858886

[48] Hao J, Tan K, Tjin S, et al. Design of a foot-pressure monitoring transducer for diabetic patients based on FBG sensors. In: Proceedings the 16th Annual Meeting of the IEEE Lasers and Electro-Optics Society; 27-28 October 2003; Tucson, USA. New York: IEEE; 2004. pp. 23-24. DOI: 10.1109/LEOS.2003.1251581

[49] Suresh R, Bhalla S, Tjin S, et al. Plantar stress monitoring using fibre Bragg grating sensors. In: Proceedings of the 9th International Symposium on Advanced Science and Technology in Experimental Mechanics; 1-6 November 2014; New Delhi, India

[50] Suresh R, Bhalla S, Hao J, et al. Development of a high resolution plantar pressure monitoring pad based on fibre Bragg grating (FBG) sensors. Technology and health care: Official journal of the European Society for Engineering and Medicine. 2015;**23**:785-794. DOI: 10.3233/THC-151038

[51] Liang T-C, Lin J-J, Guo L-Y. Plantar pressure detection with fibre Bragg gratings sensing system. Sensors. 2016;**16**:1766(13pp). DOI: 10.3390/s16101766

[52] Vilarinho D, Theodosiou A, Leitão C, et al. POFBG-embedded cork insole for plantar pressure monitoring. Sensors. 2017;**17**:2924(15pp). DOI: 10.3390/s17122924

[53] Zhang YF, Hong CY, Ahmed R, et al. A fibre Bragg grating based sensing platform fabricated by fused deposition modeling process for plantar pressure measurement. Measurement. 2017;**112**:74-79. DOI: 10.1016/j.measurement.2017.08.024

[54] Tjin S, Suresh R, Ngo N. Fibre Bragg grating based shear-force sensor: Modeling and testing. Journal of Lightwave Technology. 2017;**22**: 1728-1733. DOI: 10.1109/JLT.2004.831171

[55] Suresh R, Bhalla S, Singh C, et al. Combined application of FBG and PZT sensors for plantar pressure monitoring at low and high speed walking. Technology and Health Care. 2014;**23**(1):47-61. DOI: 10.3233/THC-140867

[56] Koulaxouzidis A, Holmes M, Roberts C, et al. A shear and vertical stress sensor for physiological measurements using fibre Bragg gratings. In: Proceedings of the 22nd Annual International Conference of the IEEE Engineering in Medicine and Biology Society (Cat. No.00CH37143); 23-28 July 2000; Chicago, USA. New York: IEEE; 2002. pp. 55-58. DOI: 10.1109/IEMBS.2000.900666

[57] Zhang Z, Tao X, Zhang H, et al. Soft fibre optic sensors for precision measurement of shear stress and pressure. IEEE Sensors Journal. 2013;**13**:1478-1482. DOI: 10.1109/JSEN.2012.2237393

[58] Chethana K, Prasad A, Omkar S, et al. Design and development of optical sensor based ground reaction force measurement platform for gait and geriatric studies. World Academy of Science, Engineering and Technology, International Journal of Mechanical, Aerospace, Industrial, Mechatronic and Manufacturing Engineering. 2015;**10**:60-64. DOI: 10.1999/1307-6892/10003366

[59] Tavares C, Domingues M, Frizera-Neto A, et al. Biaxial optical fibre sensor based in two multiplexed Bragg gratings for simultaneous shear stress and vertical pressure monitoring. In: Proceedings of the SPIE Photonics Europe; 22-26 April 2018; Strasburg,

France. Bellingham: SPIE; 2018. DOI: 10.1117/12.2306889

[60] Najafi B, Wrobel J, Grewal G, et al. Plantar temperature response to walking in diabetes with and without acute Charcot: The Charcot activity response test. Journal of Aging Research. 2012:140968(5pp). DOI: 10.1155/2012/140968

[61] Uustal H. Prosthetics and orthotics. In: Cooper G, editor. Essential Physical Medicine and Rehabilitation. United States: Humana Press; 2006. pp. 101-118. DOI: 10.1007/978-1-59745-100-0

[62] Mishra V, Singh N, Tiwari U, et al. Fibre grating sensors in medicine: Current and emerging applications. Sensors and Actuators: A. 2011;**167**:279-290. DOI: 10.1016/j.sna.2011.02.045

[63] Saggio G, Quitadamo L, Albero L. Development and evaluation of a novel low-cost sensor-based knee flexion angle measurement system. The Knee. 2014;**21**:896-901. DOI: 10.1016/j.knee.2014.04.014

[64] Roriz P, Carvalho L, Frazão O, et al. From conventional sensors to fibre optic sensors for strain and force measurements in biomechanics applications: A review. Journal of Biomechanics. 2014;**47**:1251-1261. DOI: 10.1016/j.jbiomech.2014.01.054

[65] Abro Z, Yi-Fan Z, Cheng-Yu H, et al. Development of a smart garment for monitoring body postures based on FBG and flex sensing technologies. Sensors and Actuators A: Physical. 2018;**272**: 153-160. DOI: 10.1016/j.sna.2018.01.052

[66] Mohanty L, Tjin S, Lie D, et al. Fibre grating sensor for pressure mapping during total knee arthroplasty. Sensors and Actuators: A. 2007;**135**:323-328. DOI: 10.1016/j.sna.2006.07.021

[67] Rocha R, Silva A, Carmo J, et al. FBG in PVC foils for monitoring the knee joint movement during the rehabilitation process. In: Proceedings of the 33rd Annual International Conference of the IEEE Engineering in Medicine and Biology Society; 30 August–3 September 2011; Boston, USA. New York: IEEE; 2011. pp. 458-461. DOI: 10.1109/IEMBS.2011.6090064

[68] Deluzio K, Wyss UP, Zee B, et al. Principal component models of knee kinematics and kinetics: Normal vs. pathological gait patterns. Human Movement Science. 1997;**16**:201-217. DOI: 10.1016/S0167-9457(96)00051-6

[69] Anderson D, Sidaway B. Coordination changes associated with practice of a soccer kick. Research Quarterly for Exercise and Sport. 1994;**65**:93-99. DOI: 10.1080/02701367.1994.10607603

[70] Rocha L, Correia J. Wearable sensor network for body kinematics monitoring. In: Proceedings 2006 10th IEEE International Symposium; 11-14 October 2006; Montreux, Switzerland. New York: IEEE 2007. pp. 137-138. DOI: 10.1109/ISWC.2006.286364

[71] Skelly M, Chizeck H. Real-time gait event detection for paraplegic FES walking. IEEE Transactions on Neural Systems and Rehabilitation Engineering. 2001;**9**:59-68. DOI: 10.1109/7333.918277

[72] Castro M, Abreu S, Sousa S, et al. In-shoe plantar pressures and ground reaction forces during overweight adults' overground walking. Research Quarterly for Exercise and Sport. 2014;**85**:188-197. DOI: 10.1080/02701367.2014.893055

[73] Umesh S, Padma S, Srinivas T, et al. Fibre Bragg grating goniometer for joint angle measurement. IEEE Sensors Journal. 2018;**18**:216-222. DOI: 10.1109/JSEN.2017.2770176

[74] Ren L, Song G, Conditt M, et al. Fibre Bragg grating displacement

sensor for movement measurement of tendons and ligaments. Applied Optics. 2007;**46**:6867-6871. DOI: 10.1364/AO.46.006867

[75] Al-Fakih E, Osman N, Eshraghi A, et al. The capability of fibre Bragg grating sensors to measure amputees' trans-tibial stump/socket interface pressures. Sensors. 2013;**13**:10348-10357. DOI: 10.3390/s130810348

[76] Dou P, Jia X, Suo S, et al. Pressure distribution at the stump/socket interface in transtibial amputees during walking on stairs, slope and non-flat road. Clinical Biomechanics. 2006;**21**:1067-1073. DOI: 10.1016/j.clinbiomech.2006.06.004

[77] Al-Fakih E, Osman N, Eshraghi A. Techniques for interface stress measurements within prosthetic sockets of transtibial amputees: A review of the past 50 years of research (Review). Sensors. 2016;**16**:1119(30pp). DOI: 10.3390/s16071119

[78] Osman N, Spence W, Solomonidis S, et al. The patellar tendon bar! Is it a necessary feature? Medical Engineering and Physics. 2010;**32**:760-765. DOI: 10.1016/j.medengphy.2010.04.020

[79] Buis A, Convery P. Calibration problems encountered while monitoring stump/socket interface pressures with force sensing resistors: Technique adopted to minimize inaccuracies. Prosthetic and Orthotics International. 1997;**21**:179-182. DOI: 10.3109/03093649709164552

[80] Dumbleton T, Buis A, McFadyen A, et al. Dynamic interface pressure distributions of two transtibial prosthetic socket concepts. Journal of Rehabilitation Research and Development. 2009;**46**:405-415. DOI: 10.1682/JRRD.2008.01.0015

[81] Portnoy S, Yarnitzky G, Yizhar Z, et al. Real-time patient-specific finite element analysis of internal stresses in the soft tissues of a residual limb: A new tool for prosthetic fitting. Annals of Biomedical Engineering. 2007;**35**:120-135. DOI: 10.1007/s10439-006-9208-3

[82] Moo E, Osman N, Pingguan-Murphy B, et al. Interface pressure profile analysis for patellar tendon-bearing socket and hydrostatic socket. Acta of Bioengineering and Biomechanics. 2009;**11**:37-43. Available from: http://www.ncbi.nlm.nih.gov/pubmed/20405814

[83] Dickinson A, Steer J, Worsley P. Finite element analysis of the amputated lower limb: A systematic review and recommendations. Medical Engineering and Physics. 2017;**43**:1-18. DOI: 10.1016/j.medengphy.2017.02.008

[84] Tsiokos D, Kanellos G, Papaioannou G, et al. Fibre-optic-based pressure sensing surface for skin health management in prosthetic and rehabilitation interventions. In: Hudak R, editor. Biomedical Engineering–Technical Applications in Medicine. United Kingdom: IntechOpen; 2012. pp. 245-268. DOI: 10.5772/50574

[85] Kanellos G, Papaioannou G, Tsiokos D, et al. Two dimensional polymer-embedded quasi-distributed FBG pressure sensors for biomedical applications. Optics Express. 2010;**18**:179-186. DOI: 10.1364/OE.18.000179

[86] Kanellos G, Tsiokos D, Pleros N, et al. Enhanced durability FBG-based sensor pads for biomedical applications as human-machine interface surfaces. In: Proceeding of the International Workshop on BioPhotonics; 8-10 June 2011; Parma, Italy. New York: IEEE; 2011. 3pp. DOI: 10.1109/IWBP.2011.5954848

[87] Al-Fakih E, Osman N, Adikan F, et al. Development and validation of fibre Bragg grating sensing pad

for interface pressure measurements within prosthetic sockets. IEEE Sensors Journal. 2016;**16**:965-974. DOI: 10.1109/JSEN.2015.2495323

[88] Al-Fakih E, Arifin N, Pirouzi G, et al. Optical fibre Bragg grating-instrumented silicone liner for interface pressure measurement within prosthetic sockets of lower-limb amputees. Journal of Biomedical Optics. 2017;**22**:087001(8pp). DOI: 10.1117/1.JBO.22.8.087001

[89] Paternò L, Ibrahimi M, Gruppioni E, et al. Sockets for limb prostheses: A review of existing technologies and open challenges. IEEE Transactions on Biomedical Engineering. 2018;**65**:1996-2010. DOI: 10.1109/TBME.2017.2775100

[90] Norhafizan A, Ghazilla R, Kasi V, et al. A review on lower-limb exoskeleton system for sit to stand, ascending and descending staircase motion. Applied Mechanics and Materials. 2014;**541**:1150-1155. DOI: 10.4028/www.scientific.net/AMM.541-542.1150

[91] Rocon E, Pons J, editors. Exoskeletons in Rehabilitation Robotics-Tremor Suppression. Berlin: Springer Tracts in Advanced Robotics; 2011. DOI: 10.1007/978-3-642-17659-3

[92] Chen G, Chang C, Guo Z, et al. A review of lower extremity assistive robotic exoskeletons in rehabilitation therapy. Critical Review in Biomedical Engineering. 2013;**41**:343-363. DOI: 10.1615/CritRevBiomedEng.2014010453

[93] Blank A, French J, Pehlivan A, et al. Current trends in robot-assisted upper-limb stroke rehabilitation: Promoting patient engagement in therapy. Current Physical Medicine and Rehabilitation Reports. 2014;**2**:184-195. DOI: 10.1007/s40141-014-0056-z

[94] Ayoade M, Morton L, Baillie L. Investigating the feasibility of a wireless motion capture system to aid in the rehabilitation of total knee replacement patients. In: Proceedings of the IEEE 5th International Conference on Pervasive Computing Technologies for Healthcare (PervasiveHealth) and Workshops; 23-26 May 2011; Dublin. New York: IEEE; 2011. pp. 404-407. DOI: 10.4108/icst.pervasivehealth.2011.246132

[95] Park YL, Ryu S, Black R, et al. Exoskeletal force-sensing end-effectors with embedded optical fibre-Bragg-grating sensors. IEEE Transactions on Robotics. 2009;**25**:1319-1331. DOI: 10.1109/TRO.2009.2032965

[96] Jiang L, Low K, Cost J, et al. Fibre optically sensorized multi-fingered robotic hand. In: Proceedings of the International Conference on Intelligent Robots and Systems; 28 September-2 October 2015; Hamburg, Germany. New York: IEEE; 2015. pp. 1763-1768. DOI: 10.1109/IROS.2015.7353606

[97] Fujiwara E, Miyatake D, Santos M, et al. Development of a glove-based optical fibre sensor for applications in human-robot interaction. In: Proceedings of the 8th ACM/IEEE International Conference on Human-robot Interaction (HRI); 3-6 March 2013; Tokyo, Japan. New York: IEEE; 2013. pp. 123-124. DOI: 10.1109/HRI.2013.6483532

[98] Aram K, Al-Jumaily A. Active exoskeleton control systems. State of the art. Proceeding Engineering. 2012;**41**:988-994. DOI: 10.1016/j.proeng.2012.07.273

Chapter 4

Distributed, Advanced Fiber Optic Sensors

Sanjay Kher and Manoj Kumar Saxena

Abstract

India is poised to use nuclear energy in a big way. The safety of these systems depends upon monitoring various parameters in hazardous environment like high radiation, high temperature exceeding 1000°C, and gas/coolant leakages. In this chapter, we shall dwell on basics of distributed sensing, related instrumentation, device fabrication, and actual advanced field applications. Techniques like Raman scattering, resonance response of fiber gratings, and selective absorption are employed for design, development, and fabrication of distributed sensors and devices. Raman distributed sensors with advanced data processing techniques are finding increasing applications for fire detection, coolant leak detection, and safety of large structures. The systematic investigations related to portable systems developed at the author's lab have been described. Wavelength-encoded fiber gratings are the attractive candidate for high gamma radiation dose measurements in environment such as particle accelerators, fission reactors, food processing facilities, and ITER-like installations. The basics of fiber gratings, their operational designs, and devices based on fiber gratings have been described with advanced applications like high temperature sensing, strain measurements at cryogenic temperatures, and strain in nuclear environment. Finally, novel approaches are described for distributed hazardous gas monitoring for large areas such as airports, train stations, and reactor containment buildings.

Keywords: Raman optical fiber distributed temperature sensor, fiber Bragg grating sensor, long period grating sensor, dynamic self-calibration, strain sensors for nuclear environment

1. Introduction

The discovery of lasers in the 1960s and development of low loss silica optical fiber opened a new era of fiber optic sensors. Intrinsic insensitivity to electromagnetic interference (EMI), remote detection, operational ability in hazardous environment, and potential for distributed sensing make them especially useful for monitoring large nuclear infrastructures such as coolant monitoring, reactor containment buildings, nuclear waste storage sites etc. [1–5]. Radiation tolerant fibers can be used in various configurations for distributed sensing of temperature, strain, and several other parameters avoiding the requirement of positioning many discreet sensors [6–11]. Further, the radiation sensitive fibers can be used for radiation dose monitoring for local dose deposition measurements, hot spot dose monitoring in waste storage facilities, surveillance at airports and ports of entry, etc. With the availability of lasers, fibers, and low noise detectors in the mid-IR region, it has become possible to design novel distributed sensor devices for sensing hazardous volatile compounds for homeland security especially at airports, underground metro stations, and big event areas.

A wide range of techniques such as intensity modulation, wavelength encoding, and polarization provide powerful sensing capabilities. Further, several detection techniques have been investigated for development of optical fiber-based distributed sensors. Radiation-induced absorption, scintillation, fluorescence, optically stimulated luminescence, and induced refractive index changes have been used for real-time dose measurements. Optical fiber grating [12–15]-based specialty sensors have been used for distributed strain measurements in very low temperature, very high temperature, or high radiation environment. Raman and Brillouin scattering-based techniques are used for distributed temperature measurements for fire and hot spot detection. Mid-IR and near-IR absorption measurements coupled with hollow core fibers are used for leak detection of hazardous gases. This chapter will describe the basic principles, main components, various sensing systems for advanced applications, and future potential of distributed fiber sensors.

2. What is distributed sensing?

Distributed sensing is a technique whereby one sensor cable is capable to collect data (continuous/quasi-continuous profiling) that are spatially distributed over many individual measurement points. The various modes of sensing can be understood from **Figure 1** [3, 4]. Briefly, a point sensor means monitoring a parameter at a discreet point; a quasi-distributed sensor system involves an arrangement of a finite number of discreet sensors as a linear array, while in fully distributed sensing mode, the measuring parameter of interest is monitored continuously along the fiber path, providing a spatial mapping of the parameter along fiber.

In conventional sensing, say for temperature, an individual sensor such as a thermocouple or platinum probe is needed for each point of interest whereas distributed sensing addresses many points simultaneously along with their spatial location [5]. With proper design architecture, it can contribute to enhanced safety and security by providing early warnings of gas and coolant leakages, structural cracks, onset of fire, hot spot detection in pipelines, radiation leaks, etc. Two techniques such as optical time-domain reflectometry (OTDR) [4, 16–23] and

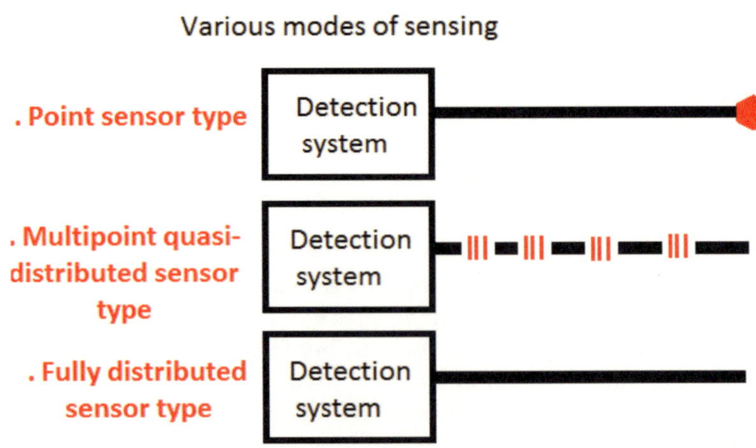

Figure 1.
Various modes of sensing: point, multipoint quasi-distributed, and fully distributed.

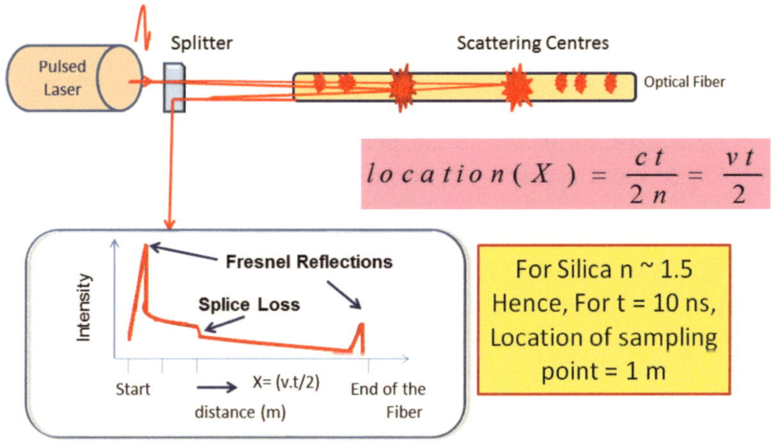

Figure 2.
Schematic diagram of OTDR.

wavelength division multiplexing (WDM) are generally used for distributed sensing. In OTDR (**Figure 2**), a pulsed laser is coupled to an optical fiber through a directional coupler/splitter. The backscattered light originating from density and composition variation is monitored continuously in time. The spatial location of an event is determined from time of flight measurements, that is, the device calculates the distance of the measuring point based on the time it takes for the reflected light to return.

For example, if the backscattered light is detected after 10 ns from the starting point, it is set to originate from 1 m distance from origin of fiber. This can be easily calculated from OTDR equation $X = ct/2n$, where X is the distance from origin (start of fiber taken as zero time), t is the time of event detection, c (3×10^8 m/s) is the velocity of light in the vacuum, and n is the refractive index of fiber for wavelength of operation. If we use a sensing fiber with core refractive index (n) of 1.5 and wish to measure distance (X) traveled after $t = 10$ ns, then it is easy to calculate that $X = 1.0$ m by putting the values in OTDR equation [21, 23].

In wavelength multiplexing, a device such as Bragg grating is used to encode a series of resonant wavelengths in the fibers [12–15]. The wavelengths in turn are monitored by wavelength interrogator. The resonant wavelengths are affected by measuring parameters and are thus monitored in a quasi-distributed manner.

3. All fiber Raman optical fiber distributed temperature sensor with dynamic self-calibration

Temperature sensors are ubiquitous devices that permeate our daily lives. Many areas of temperature measurements require a large area of coverage with high localization accuracy. Raman optical fiber-based distributed temperature sensors (ROFDTSs) are equipped with the ability of providing temperature values as a continuous function of distance along the fiber. In an ROFDTS, every bit of fiber works as a sensing element as well as data transmitting medium, to substitute the role played by several point sensors, thus allowing reduced sensor network cost. ROFDTSs have attracted the attention as a means of temperature monitoring and fire detection in power cables, long pipelines, bore holes, tunnels, and critical installations like oil wells, refineries, induction furnaces, and process control industries. The basic principle of temperature measurement using ROFDTS involves

Raman scattering [10] in conjunction with OTDR. The ratio of Raman anti-Stokes (AS) and Stokes (St) intensities is used for determination of unknown temperature. The AS signal is strongly dependent on temperature, while the Stokes signal is slightly dependent on temperature. Based on time of flight and intensity of Stokes and anti-Stokes signals, location and temperature information can be retrieved. The backscattered light has many spectral components as shown in **Figure 3** [24]. For temperature measurements, Raman components are analyzed.

The OTDR principle allows estimation of the location of hot zone whereas Raman scattering permits measurement of temperature of the hot zone. Sensing fiber is coupled to short interrogating laser pulses, and backscattered AS and St components are monitored for signal changes. Unknown temperature of hot zones can be estimated from the ratio (R) of AS and St using the following expression [11]

$$R = \frac{I_{as}}{I_s} = \left(\frac{\lambda_s}{\lambda_{as}}\right)^4 \exp\left(-\frac{hc\nabla}{kT}\right) \tag{1}$$

where λ_s and λ_{as} are the Stokes and anti-Stokes optical signal wavelengths, ∇ is their wave number separation from the pump laser wavelength, h is Planck's constant, c is the velocity of light, and k is Boltzmann's constant. AS is the main signal which carries the signature of temperature variation whereas St provides reference and eliminates a number of effects common to both the signals. One can simplify Eq. (1) by replacing known terms by B where,

$$B = \frac{hc\nabla}{k} \tag{2}$$

Since values of h, c, and k are known, the numerical value of B is found to be 631.3 for silica fiber having ∇ = 440 cm^{-1}.

One can simplify the profile analysis by referencing the ratio profile at unknown temperature to the ratio value at known temperature of a pre-selected calibration zone of fiber. The temperature of a given zone T (°C) is then given by the following expression [25].

$$T(^\circ\mathrm{C}) = B.\left[\frac{1}{\left(B.\frac{1}{\theta} - \ln R_T + \ln R_\theta\right)}\right] - 273 \tag{3}$$

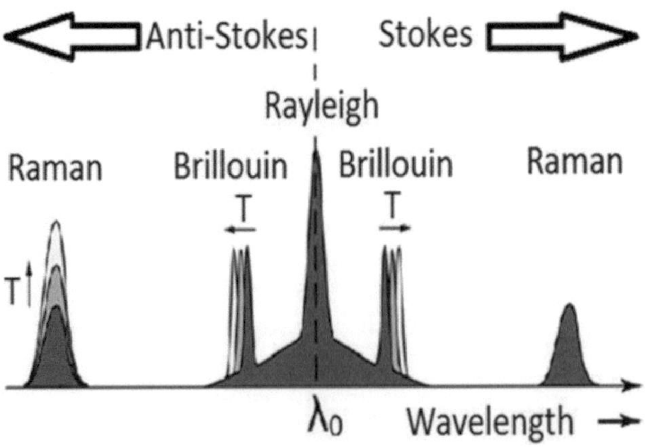

Figure 3.
Backscattered laser light from optical fiber in the case of Raman and Brillouin scattering (source: https://www.google.co.in/search?q=BACKSCATTERED+LASER+LIGHT&source=lnms&tbm=isch&sa=X&ved=0ahUKEwiCvf-iq9jcAhWKT3oKHdJ4CjYQ_AUICigB&biw=1093&bih=530) (images) [24].

Here, the calibration zone is kept at some known absolute room temperature (Θ). Eq. (3) can be deduced after taking the quotient of the ratio profile at unknown temperature (R_T) for an arbitrary zone and the ratio value at the calibration zone (R_Θ) and solving it for T. Parameter (R_Θ) is the ratio value of AS to St signal (AS/St) for the calibration zone (at temperature Θ) of length 1 m chosen from sensing fiber at the laser end. In Eq. (3), parameters B and Θ are known. Therefore, Eq. (3) will yield temperature profile (T in °C) for complete fiber length, provided that profiles of R_T and value of R_θ are available.

Figure 4 shows the block diagram of Raman optical fiber distributed temperature sensor.

Further, **Figure 5** gives an idea of averaged anti-Stokes Raman signal for a 2.5 m long zone heated by a proportional intergral derivative (PID) controlled heating oven.

To determine the unknown temperature profile with certain accuracy for complete fiber length by using Eq. (1), appropriate measures are to be devised and implemented to address several error-causing issues. These issues are described as below [21, 23]. The author's laboratory has successfully solved these issues and designed a field portable unit.

3.1 Issue no. 1

The first issue is the difference in theoretical and experimental values of the ratio (R) at various temperature values. For example, at room temperature (25°C, say), the theoretical and experimental values of R are 0.1693 and 0.55, respectively. At 50°C, the theoretical and experimental values of R are found to be 0.1995 and 0.658, respectively. On the other hand, the theoretical and experimental values are 0.2415 and 0.8279, respectively, at 85°C. The reason for this difference is explained below [23].

At 25°C (for example), obtaining a theoretical value of 0.1693 for R requires that the optoelectronic conversion using photomultiplier tube (e.g. PMT-R5108, Hamamatsu) detectors, the beam splitting, and the subsequent light coupling into AS and St detectors are in such a way that the relation St = 5.906 × AS is maintained for backscattered AS and St signals while traveling the path from fiber to the final stage of detection. However, due to nonideal behavior of various optical components in

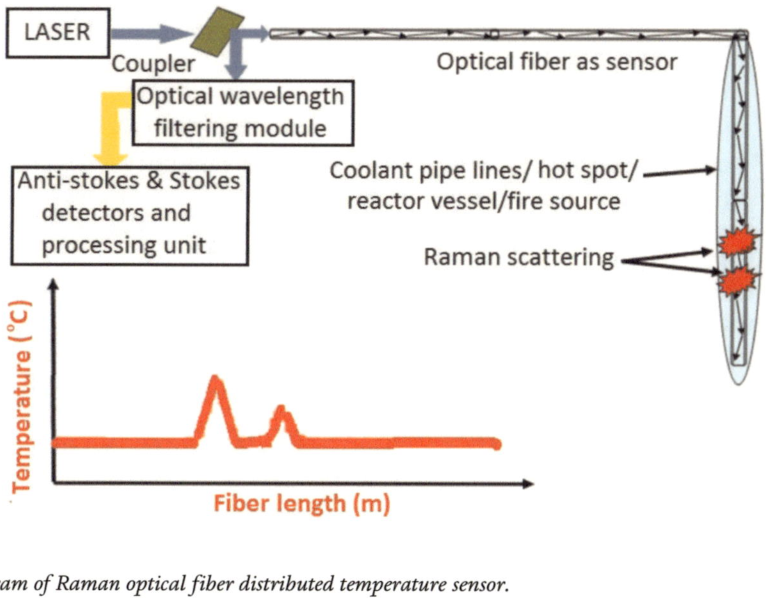

Figure 4.
Block diagram of Raman optical fiber distributed temperature sensor.

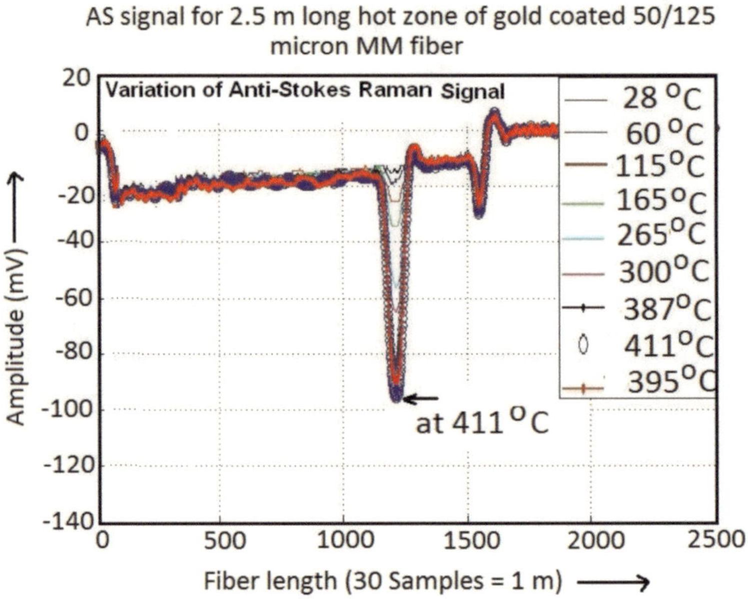

Figure 5.
The anti-Stokes Raman signal profile at various temperatures.

the path and band nature of AS and St signals, the above relation does not hold. The relation gets deteriorated at every stage in the path. For example, the cathode radiant sensitivity of a PMT for AS wavelength (1018 nm) and St wavelength (1109 nm) is 0.95 mA/W and 0.2mA/W, respectively, which causes St current to be approximately 5 times less compared to AS current. Nonideal performance of beam splitters and optical filters also does not support the above ideal relation. As a result, the cumulative effect of various components makes experimental values of R to be different from the theoretical one.

Direct use of experimental values of R in Eq. (1) will yield highly erroneous and unacceptable temperature profile (T). Hence, Eq. (1) needs to be modified to obtain correct values. Modification is done by referencing the experimentally obtained ratio values with respect to the ratio value at some known temperature of calibration zone which is chosen from sensing fiber itself.

3.2 Issue no. 2

The second issue is the nonidentical fiber attenuation along the fiber length for Raman AS and St signals due to difference in their wavelengths [9, 11]. In a typical system using 1064 nm excitation laser, the difference between two wavelengths is ~90 nm. The lower optical wavelength signal (AS) experiences higher attenuation in comparison to higher optical wavelength signal (St) while traveling in sensing optical fiber. This attenuation difference results in an unwanted downward slope in ratio (R) profile and finally in unknown temperature (T) profile with respect to fiber length. It may be noted that downward slope in ratio (R) profile causes additional errors in unknown temperature (T) profile of fiber and should be corrected.

3.3 Issue no. 3

While de-noising Raman AS and St signals for better signal-to-noise ratio (SNR), conventional finite impulse response/infinite impulse response (FIR/IIR)-based

Fourier filtering causes spatial inaccuracy in locating the hot-zones which in turn yields erroneous information about the location of hot zones [18].

3.4 Issue no. 4

The amplitude of AS and St signals varies with time due to slow variations/drifts in laser power and laser-fiber coupling. Also, the temperature of calibration zone itself may change unless it is controlled by a dedicated setup. Therefore, any previously stored reference values of AS and St signals and calibration zone temperature can no longer be used as a reference for temperature measurement at a later stage.

3.5 Approaches to solve the problematic issues

Stoddart et al. [20] proposed to use Rayleigh instead of St from the backscattered spectrum to avoid the temperature measurement error in hydrogen-rich environments due to differential attenuation caused by the optical fiber for AS and St signal wavelengths. This resulted in better results but could not eliminate the error caused by the differential attenuation completely. The dual-ended (DE) configuration [26] (i.e. both ends of sensing fiber are connected to ROFDTS unit) and dual laser source schemes [27, 28] have also been proposed to take care of the difference in attenuation between AS and St. These schemes have resulted in improvements but add complexity and need double length of fiber, extra distributed temperature sensor (DTS) with an optical switch, and two costly lasers. A correction method to take care of the difference in attenuation for AS and St signals has been proposed with only one light source and one light detector but requires attachment of a carefully designed reflective mirror at the far fiber-end of the sensing fiber [29]. Recently, a more sophisticated correction technique [30] based on detection of AS signal alone in combination with DE configuration has been investigated. ROFDTSs based on the above schemes are important and to a certain extent become mandatory in situations where sensing fiber is exposed to the severe radiation environment or hydrogen darkening in oil wells. Requirements for less demanding situations like temperature measurement in steam pipelines of turbines, electrical cables and temperature profiling of big buildings, gas pipelines and mines etc. can be met by the technique based on digital signal processing.

In order to address the above issues satisfactorily, a discrete wavelet transform (DWT)-based dynamic self-calibration and de-noising technique is used and implemented by the authors as given in detail [23]. Briefly, wavelets are mathematical functions that can be used to segregate data into various frequency components. Each component can then be studied with a resolution matched to its scale. In DWT, a signal may be represented by its low frequency component and its high frequency component.

The DWT-based technique is simpler, more automatic, and provides a single solution to address all the above issues simultaneously. The DWT technique takes care of the difference in optical attenuation for AS and St signals by using their trend and also de-noises the AS and St signals while preserving spatial locations of peaks. Also, this technique requires just 1 m long calibration zone which is much less than the 100 m required in the previous technique. Moreover, the dynamic measurement of calibration zone's temperature eliminates the requirement of keeping the calibration zone at a constant temperature, and thus, complicated heating arrangement is avoided. Actual wavelet transform-based processed signal profile is shown in **Figure 6**. **Table 1** presents the comparison of error in temperature measurement at various zones using Eq. (3) with unprocessed and processed Raman signals. Both absolute errors and percentage errors (in brackets) are

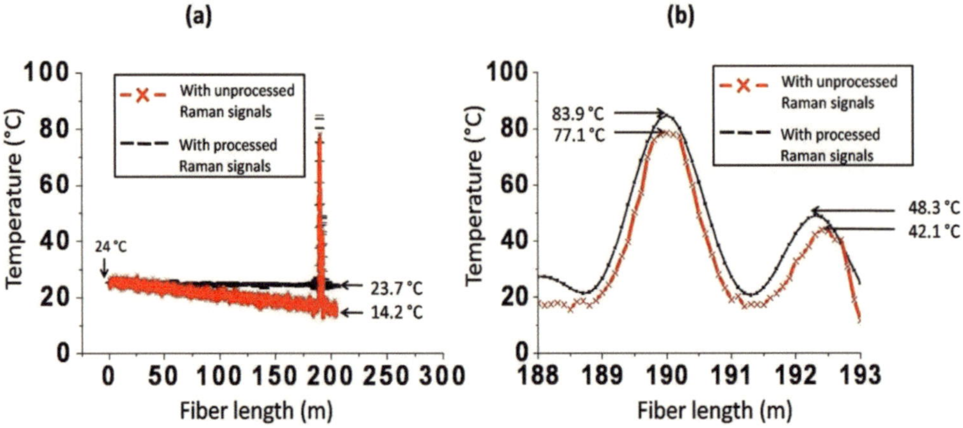

Figure 6.
Distributed temperature profile with processed (black color) and unprocessed (red color) Raman signals: (a) view for complete fiber length and (b) zoomed view for hot zones.

Zone (location)	Reference temperature	Measured temperature	
		Unprocessed	Processed
Start of fiber (location: 0 m)	24.5°C	24°C	24°C
	Error	−0.5°C (−2.04%)	−0.5°C (−2.04%)
Hot zone-1 (location: 190 m)	85°C	77.1°C	83.9 °C
	Error	−7.9 °C (−9.29%)	−1.1 °C (−1.29%)
Hot zone-2 (location: 192.4 m)	50°C	42.1°C	48.3°C
	Error	−7.9°C (−15.8%)	−2.4°C (−3.4%)
End of fiber (location: 205 m)	25°C	14.2 °C	23.7°C
	Error	−10.8°C (−43.2%)	−1.3°C (−5.2%)

Table 1.
Comparison of error in temperature measurement at various zones with unprocessed and processed Raman signals.

reported to appreciate the improvement achieved after processing of Raman signals.

Fiber Sensors Lab., Raja Ramanna Centre for Advanced Technology (RRCAT), Indore, India has developed a Raman scattering-based OFDTS [23] with the following specifications, and the developed OFDTS is capable of working in high accelerating voltage (1.5 MV), magnetic field (1.5 T), and bremsstrahlung radiation present in accelerator systems.

(a) Temperature range: 25–300°C, (b) temperature resolution: 3°C, (c) spatial resolution: 1 m (over a length of 500 m); can be improved to few cm with special fiber-laying techniques, (d) distance (dynamic range for distance covered): 500 m, (e) fire alarm: audio-visual alarms can be generated, and (f) gamma field operation; can operate up to a gamma dose of 1 MGy.

For more ruggedness and field deployability, an all-fiber ROFDTS scheme is desirable. The schematic design of one such scheme is depicted in **Figure 7**. Recently, a distributed sensor using a superconducting nanowire single photon detector and chalcogenide fiber has been proposed. This scheme has the potential to

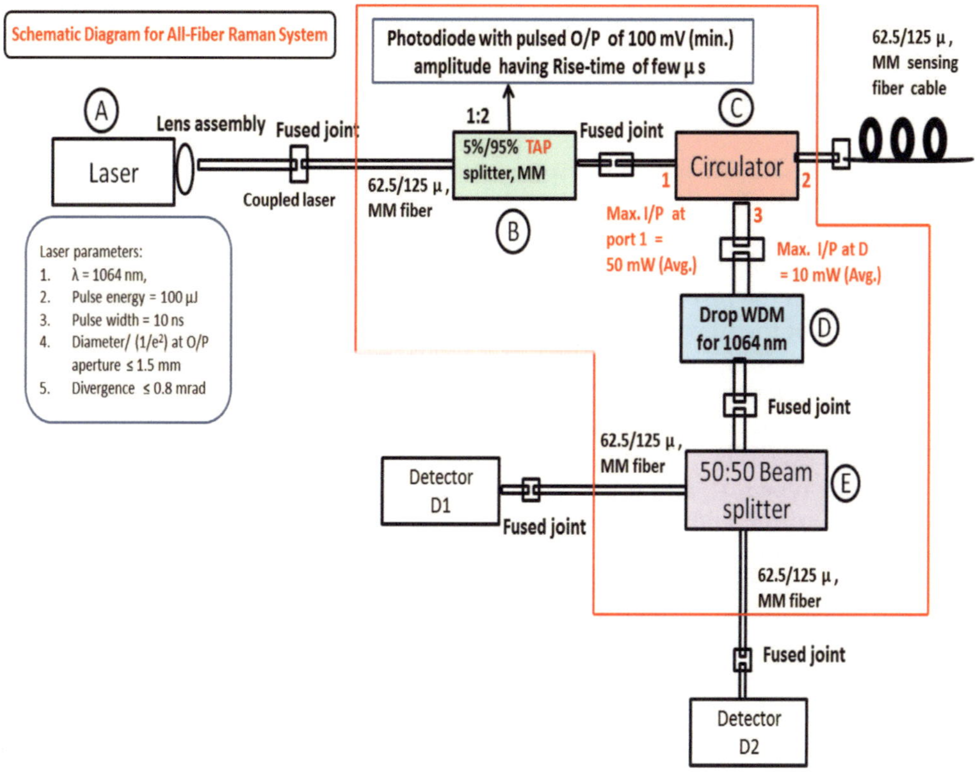

Figure 7.
Schematic diagram of an all-fiber-based ROFDTS scheme.

offer sub-centimeter spatial resolution sensor, below 1°C temperature resolution over a distance of few hundreds of meters.

4. Quasi-distributed sensors for temperature and strain measurements

The fiber Bragg gratings (FBG) were first written by Hill et al. [12] who discovered the breakthrough phenomena of photosensitivity in optical fiber. As a result of this development, FBG-based strain and temperature sensors came into existence. The method of writing FBG in sensing fiber's section involves creation of periodic modulation of fiber core's refractive index. The refractive index is modulated by spatial pattern of ultraviolet (UV) light between 240 and 260 nm. The periodic structure in fiber's core can be created by phase mask method [13, 15]. A particular pattern in a particular segment of fiber will correspond to a specific Bragg reflection wavelength. The multiple gratings can be fabricated by using a specific phase mask with different initial Bragg wavelength gratings in the same fiber causing creation of several point sensors in a single sensing fiber. Such FBG-based sensors are quasi-distributed temperature sensors where temperature sensing by fiber is possible only where grating was created.

According to Bragg's law, when a broad band light is injected into the optical fiber consisting of FBG sensors, a specific wavelength of light is reflected by FBG [15]. The Bragg wavelength is determined by the product of effective refractive index (n_{eff}) of the grating and the grating period (Λ) (also called pitch length) as given by the following equation:

$$\lambda_B = 2n_{eff}\Lambda \tag{4}$$

Figure 8 depicts the basic principle of FBG reflection spectra and interrogation technique [12–15]. The Bragg wavelength depends on grating period of FBG and the refractive property of optical fiber.

It is clear from Eq. (4) that any change in the pitch length or refractive index will induce a shift in the resonant wavelength. Consequently, temperature, strain or deformations of the fiber can be monitored by the corresponding resonant wavelength shift. The Bragg wavelength is strain- and temperature-dependent through physical elongation or thermal change of the sensor and through the change in the fiber refractive index due to photoelastic and thermo-optic effects.

There are essentially three types of gratings which vary in photosensitivity. They are known as type I, II, and IIA with the details of each type given in [15]. Type I gratings are written with moderate intensities and exhibit an index grating right across the core. Type II gratings can be written with much higher intensities within very short times, often with a single nanosecond pulse from an excimer laser (single shot damage gratings). Type IIA gratings are regenerated gratings (RGs) specifically designed for high temperature operation. In addition, there are different physical types of gratings such as long period gratings (LPGs), chirped gratings, tilted (blazed) gratings, and micro-structured FBGs. Typical temperature sensitivity of FBG is 10 pm/°C (at 1550 nm in standard silica-based single mode fiber) and strain sensitivity is 1.2 pm/micro-strain [13].

A strong point of FBGs is their capability of multiplexing in wavelength that enable multiple points or quasi-distributed sensing. The schematic diagram of the distributed FBG sensor is shown in **Figure 9**.

There have been significant developments in two of the areas that have constrained the progress of fiber grating technology. Firstly, the issue of temperature and strain isolation has been overcome by using various techniques reported in the literature, from simply having collocated sensors that are exposed to the same temperature fluctuations to isolate stress and strain, to more complex methods, such as using tilted or chirped gratings to distinguish between the different measurands. Secondly, with improved data processing methods, simpler interrogation techniques are being utilized such that the optical signal can easily be transposed into the electrical domain, allowing the optical networks to be interfaced seamlessly with electronic systems. In addition, the production of FBGs has improved significantly through draw tower processes and automated manufacturing. One of the main advantages of FBG sensors is their ability to be easily

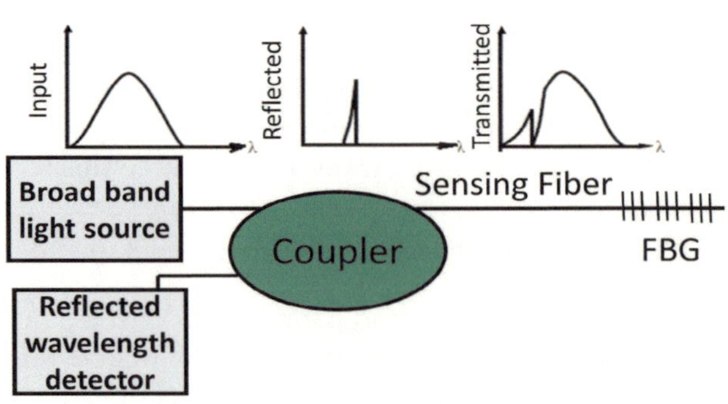

Figure 8.
Schematic representation of FBG sensor.

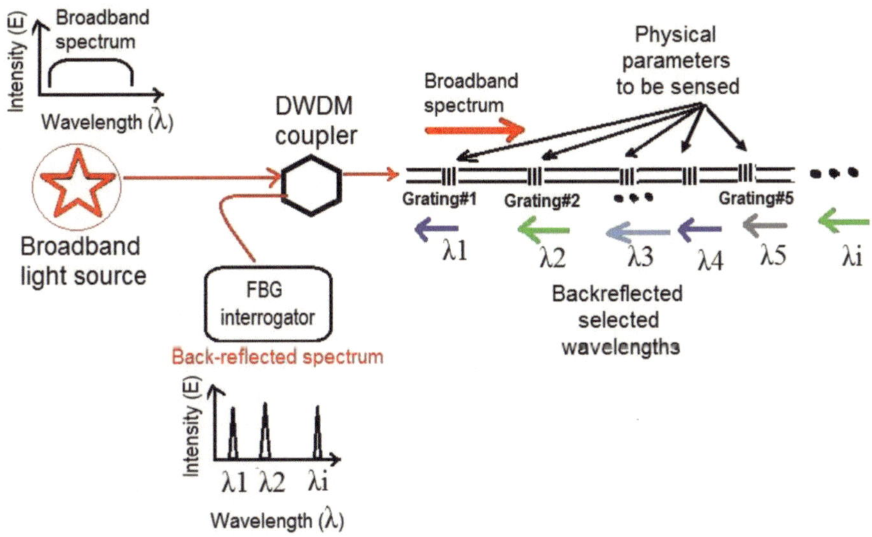

Figure 9.
Schematic diagram of the distributed FBG sensor.

multiplexed through time division multiplexing (TDM) and wavelength division multiplexing (WDM). The information from each sensor must be separated and interpreted, which requires an interrogator system to interrogate many FBGs connected in series.

A series of wavelength-encoded FBGs are used for quasi-distributed sensing applications. Several groups developed quasi-distributed FBG sensors for temperature and strain monitoring [31–33]. Central Glass & Ceramic Research Institute (CGCRI), Kolkata, India recently developed specially packaged FBGs for strain/force monitoring of electric railway engine pantographs. However, very special packaging or specialty fibers were required for advanced applications such as strain monitoring at high temperature, strain monitoring at cryogenic temperature, and very high temperature monitoring exceeding 1000°C. FBG strain sensor for health monitoring of structure at 600°C have also been developed.

5. Ultra-high temperature distributed sensors

The dynamic range of ROFDTS is restricted by coating on the optical fiber. Polyimide coatings can permit measurement up to 350°C while the gold coating may allow the measurement up to 600°C. Beyond this, distributed sensing is possible by specialized gratings made in specialized fibers. For ultra-high temperature sensing, type II-IR gratings in silica optical fiber withstand a temperature of up to 1000°C, which are usually fabricated by using a femto-second laser with power density near the damage threshold of the fiber glass. These gratings however have disadvantages as sensing elements because of asymmetric reflection spectrum and a large spectral width of more than 0.6 nm. These create problems during distributed sensing. Gratings written on a different host material, namely sapphire gratings, can be used as a temperature-sensing probe up to 1900°C. However, the material and mode mismatch with normal silica-based optical fiber and high cost of fabrication restricts its use in distributed sensing. Identification of structural changes on a molecular scale involved with the formation of a new type of FBG named regenerated Bragg grating based on annealing a conventional Type-I FBG to create a new, more robust one seems to be a promising candidate to achieve better sustainability at high temperature. However, in standard photosensitive silica fiber, RG gratings were found to be stable only below 950°C.

CGCRI, Kolkata, India in collaboration with RRCAT, Indore, India has taken up the development of a new glass composition-based photosensitive fiber to fabricate RG for temperature ~1400°C. The fiber is likely to be based on yttrium-stabilized zirconia-calcium-alumina-phospho silica glass. The motivation of the choice of such kind of multi-material glass-based optical fiber is to increase the photosensitivity along with thermal stability of fabricated RG. The regeneration takes place near the fiber glass transition temperature, in which the transformation of the glass from monoclinic structure to tetragonal structure occurs. The ultra-high temperature sustainability of RG will be evaluated for the special composition through material study, which definitely is not achievable in a standard germano-silicate fiber. This work is expected to provide a new degree of freedom in the design of optical fiber sensor for ultra-high temperature sensing. This will open up opportunities in sectors such as power plants, turbines, combustion, and aerospace engineering where often the environments are too harsh for existing FBG sensor technology and will offer a new degree of freedom in the design of optical fiber sensors.

6. Strain sensors for nuclear environment

6.1 Wavelength encoded strain sensors

A long period fiber grating sensor in photonic crystal fiber with a strain sensitivity of −2.0 pm/με and negligible temperature sensitivity is fabricated by use of CO_2 laser beam. Such a strain sensor can effectively reduce the cross sensitivity between strain and temperature. Due to single material (pure silica) construction, they have been shown to be resistant to nuclear radiation and are thus useful for applications in secondary loops of nuclear reactors. The authors' lab has designed and developed such sensor devices.

6.2 Design principle

Photonic crystal fibers (PCFs) also known as holey fibers are a new class of optical fibers that have attracted intense scientific research during past few years. Typically, these fibers incorporate a number of air holes that run along the length of the fiber, and the size, shape, and distribution of the holes can be designed to achieve various novel wave-guiding properties that may not be possible in conventional fibers. Various PCFs have been demonstrated so far that exhibit remarkable properties such as endlessly single mode fiber, large mode area, and highly nonlinear performance. Temperature-insensitive long period gratings have attracted much attention because of their potential applications in achieving stable optical filters and gain flatteners as well as in realizing temperature-insensitive sensors for industrial and nuclear applications. Conventional fibers contain at least two different glasses, each with a different thermal expansion coefficient, thereby giving rise to high temperature sensitivity. PCFs are virtually insensitive to temperature because they are made of only one material (and air hole). This property can be used to obtain temperature-insensitive PCF-based devices. Long period gratings (LPGs) in PCF fibers have not yet been reported in India. Besides, the effect of high nuclear radiation on such PCF-based grating sensors has not been reported by any group to the best of our knowledge.

DOI: http://dx.doi.org/10.5772/intechopen.83622

6.3 Theory

An LPG is formed by introducing periodic modulation of the refractive index along a single mode fiber. Such a grating induces light coupling from the fundamental guided mode to co-propagating cladding modes at discreet resonant wavelengths. LPGs in conventional fibers have been extensively used as band rejection filters, gain flattening filters, tunable couplers, and sensors. In general, as fiber devices and sensing elements, LPGs offer low back reflection, insensitivity to electromagnetic interference, and low insertion loss and cost effectiveness. For a long period grating with periodicity Λ, the wavelength $\lambda^{(m)}$ at which mode coupling occurs is given by

$$\lambda^{(m)} = \left(n_{eff} - n_{cl,\, m} \right)\Lambda \quad (5)$$

where n_{eff} is the effective refractive index of the propagating core mode at wavelength λ, and $n_{cl,m}$ is the effective refractive index of the mth cladding mode. The variation in the grating period and modal effective indices due to strain and temperature causes the coupling wavelength to shift. This spectral shift is distinct for each loss band and is a function of the order of corresponding cladding mode.

The axial strain sensitivity of LPGs may be examined by expanding Eq. (5) to yield

$$\frac{d\lambda}{d\varepsilon} = \frac{d\lambda}{d\left(\delta n_{eff}\right)} \left(\frac{dn_{eff}}{d\varepsilon} - \frac{dn_{cl}}{d\varepsilon} \right) + \Lambda \frac{d\lambda}{d\Lambda} \quad (6)$$

where δ_{neff} = $(n_{eff} - n_{cl})$ is the differential effective index; ordinal m has been dropped for the sake of simplicity. The two terms on the right side can be divided into material (first term) and waveguide (second term) contributions. The temperature sensitivity of LPG grating is given by

$$\frac{d\lambda}{d\varepsilon} = \frac{d\lambda}{d\left(\delta n_{eff}\right)} \left(\frac{dn_{eff}}{dT} - \frac{dn_{cl}}{dT} \right) + \Lambda \frac{d\lambda}{d\Lambda} \frac{1}{L} \frac{dL}{dT} \quad (7)$$

where λ is the central wavelength of the attenuation band, T is the temperature, L is the length of the LPG, and Λ is the period of the LPG. For standard long period gratings with periodicity of hundreds of micrometers, the material effect dominates the waveguide contribution. Hence, only the first term in Eqs. (6) and (7) is considered for evaluation of sensitivity. For photonic crystal fibers which are single material fibers, the first term in Eq. (7) becomes negligible, resulting in very low temperature sensitivity. This term is an order smaller than that of B-Ge-doped photosensitive fiber. This opens-up the field for PCF-based temperature-insensitive sensors.

6.4 Device designs

Inscription of LPGs has been demonstrated using various techniques such as UV treatment, heat treatment with a CO_2 laser, or by applying mechanical pressure. Formation of LPG in pure-silica core PCF fibers is not straightforward because there is no photosensitivity provided by Ge-O_2 vacancy defect centers. The LPGs in PCF are primarily formed due to modification of glass structure. However, any geometrical deformation results in flaws or cracks that result in fracture of the fiber, and therefore, LPGs in PCF require high precision systems. Our fully automated CO_2 laser-based grating writing system can set the grating period in the range of

200–800 μm with a precision of 1 μm while laser intensity can be stabilized within ±5%. **Figure 10** shows the schematic diagram of our grating writing system. The fiber is exposed to CO_2 laser for a predetermined period and the beam is scanned repeatedly over the fiber until grating of sufficient strength is formed. This operation is performed through an AutoCAD program in which the period and length of the grating are selected as per the design requirement. This method is more accurate and free from vibration related uncertainties in the grating period. The spectral response is recorded using an optical spectrum analyzer (OSA) (86142B, Agilent) which is connected to LPG through patch-cords as shown in **Figure 10**.

During application of LPG-based strain sensors, one of the main difficulties is the cross sensitivity between strain and the temperature [34]. The common methods for cross sensitivity reduction are using temperature compensation and simultaneous strain and temperature measurement. Conventional fibers contain at least two different glasses, each with a different thermal expansion coefficient, thereby giving rise to high temperature sensitivity. By use of the CO_2 laser method, an LPG sensor with strain sensitivity of −0.45 pm/με and a temperature sensitivity of 59.0 pm/°C was written in corning SMF-28 fiber 2. Another LPG with a strain sensitivity of −0.19 pm/με and a temperature sensitivity of 10.9 pm/°C was described in PCF fiber. In this paper, we present a LPG-PCF sensor fabricated in ESM-PCF with a high strain sensitivity (−2.0 pm/με) and negligible temperature sensitivity.

For the preparation of LPFG in an endless-single-mode photonic crystal fiber (ESM-PCF), both ends of the PCF are fusion spliced to SMFs [34]. The loss for each splice is about 0.74 dB. An X-Y scanning CO_2 laser is used for the fabrication of LPGs in the ESM-PCF. The CO_2 laser operates at a frequency of 2 kHz and has a maximum power of 10 W. The laser power is controlled by the mark-speed of the laser pulses. The typical grating length and period in our experiment is 23.4 mm and 450 μm, respectively. **Figure 11** shows the transmission characteristics of a LPG fabricated on an ESM-PCF. Attenuation bands in the range of 1300–1700 nm have been investigated by an Optical spectrum analyzer.

The device has been tested on a standard strain calibration platform. **Figure 12** shows the strain-dependent wavelength shift of the fabricated device.

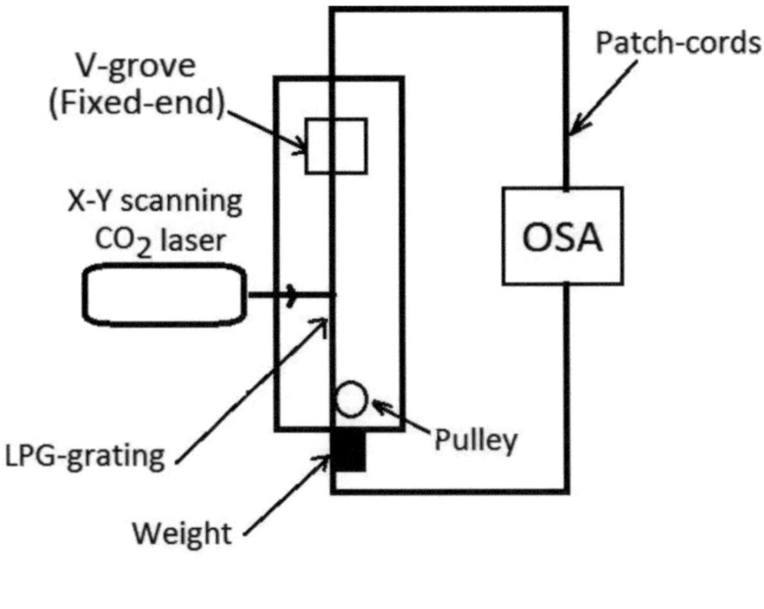

Figure 10.
Schematic diagram of long period grating fabrication set-up [34].

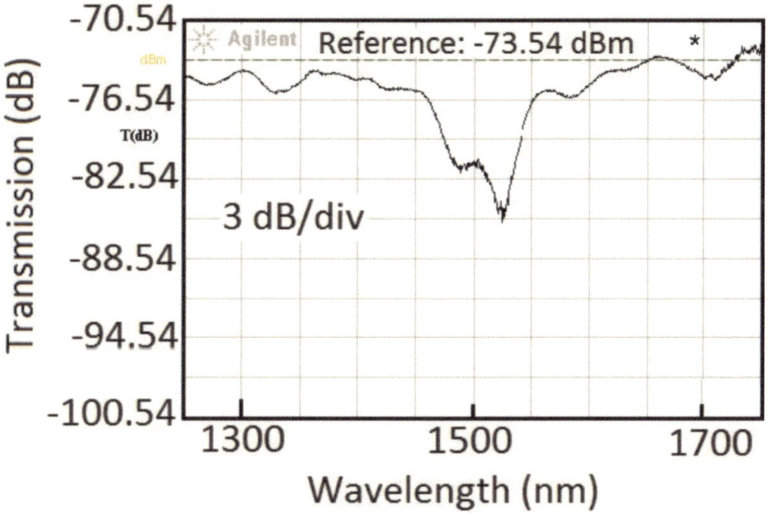

Figure 11.
Transmission characteristics of a LPG fabricated on an ESM-PCF with a period of 450 μm [34].

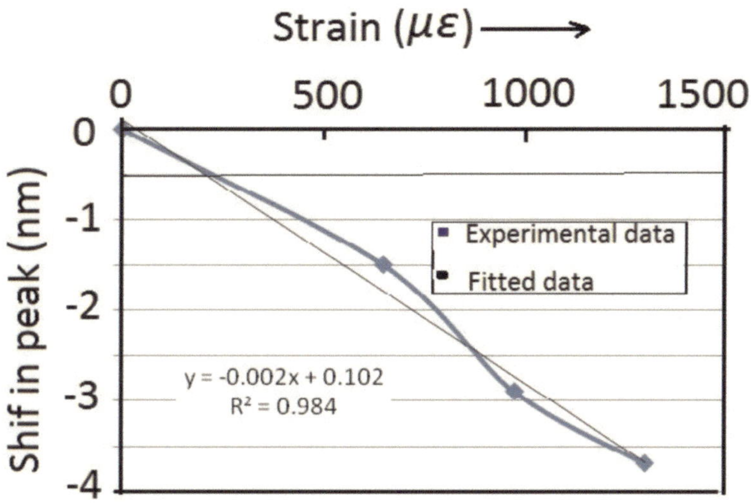

Figure 12.
Strain-dependent wavelength shift of LPG [34].

6.5 Strain sensors for cryogenic environment

The tendency of superconducting magnet coils to quench prematurely, at relatively low fractions of the critical current, or to exhibit training behavior, is often attributed to mechanical issues. Knowledge of stress, strain, and displacement of the windings is therefore central to the design of the superconducting magnet. The resistive foil strain gauge has remained the device most commonly used for measuring the strain on cryogenic structures. The nonlinear thermal apparent strains and measurement sensitivity to electromagnetic noise remain the most significant limitations to its successful implementation. FBG sensor has a number of distinct advantages over other sensors, such as EMI immunity, high sensitivity, and compact size. Furthermore, the wavelength-encoded nature allows the distributed sensing of strain. Fiber Bragg gratings are used to monitor temperature and strain in engineering structures; to date, however, their use has been limited to ambient and high temperatures, typically in the range of 273–773 K.

There exist very few published reports on FBG strain sensors that have been functional at liquid nitrogen temperature. Zhang et al. have reported one such FBG sensor used to strain sensing at 77 K and used in high temperature superconducting magnet [29]. To date, an FBG sensor used to strain sensing at 4.2 K and used in low temperature superconducting magnet has not been reported. RRCAT, Indore is also working on FBG strain and temperature sensors for cryogenic applications. The sensors show linearity in strain range from 50 to 500 micro-strain at liquid nitrogen temperature.

6.5.1 Nuclear radiation sensors

Optical fibers offer a unique capability for remote monitoring of radiation in hazardous locations such as nuclear reactors and waste storage sites. Increase of attenuation, luminescence, and radiation-induced index change have been used to design dose sensors for dose ranges up to 100 kGy. The attenuation-based sensors based on specialty doped fibers reach a saturation level above 10 kGy. To overcome this limitation, alternative techniques such as changes in fiber gratings are explored. The wavelength-encoded operation of fiber gratings can solve many measurement problems such as radiation-induced broadband transmission loss in optical fibers, source fluctuation, etc. Most Bragg grating-based sensors, reported till date, are either less sensitive or reach a saturation level near 50–150 kGy depending on the composition and grating writing technique [29–33]. Recent publications have reported measurements only up to 100 kGy. The authors Henchel et al. [35] used specialty chiral gratings and reported measurements up to 100 kGy. However, the mode orders and fiber composition in sensitive gratings were not known. Rego et al. [36] have performed gamma dose measurements on arc-induced long period fiber gratings up to 500 kGy but found no measurable shift in the resonance wavelength. Gusarov et al. [16, 17, 37] have conducted high dose measurements on FBGs but did not find high sensitivity. We have discovered sensitive gratings in commercially available single mode fibers with known composition and mode orders [38, 39]. Our results and approach are described. These are believed to be the first studies of CO_2 written long period gratings up to 1 MGy.

6.5.2 Optical fiber composition optimization for high gamma dose and temperature sensing applications

Following requirements explain the need for novel radiation dose sensors:

a. Measurement of precise dose delivery is very crucial for treatment of cancer (40–50 Gy in about 20 sittings).

b. In the case of gamma source misplacement by universities or hospitals, it is important that state-of-the-art sensors away from the source are required.

c. Accelerators, fusion reactors, nuclear waste sites, and accidental leaks in reactors all require a sensitive, large area but remote dose sensors. Typically, the dose in various conditions and installations are: Tokamak Fusion reactor system, Japan (behind coils: 2 kGy, behind tiles: 200 MGy, 1.1 m behind port plug: 15 Gy). For space-based systems, total 10 year dose is around 100 kGy.

To cover a broad dose range from few Gy to 1 MGy, novel sensor systems like gratings are desirable. For most fibers, the increase in attenuation with dose saturates near few kGy which is accumulated within a relatively short time at certain critical locations and so they need to be replaced frequently. Even space-based systems are qualified for a dose up to 100 kGy.

6.5.3 Novel devices, fabrication technology, and testing for high radiation dose detection

Specialty doped fibers are required to measure high dose gamma radiation. These fibers should have negligible radiation-induced attenuation in IR but should show high index changes upon irradiation. Wavelength encoded fiber gratings are attractive candidates for high level gamma dose measurements in nuclear environment. This paper explains for the first time how arc-induced long period fiber gratings can be optimally designed for gamma dose measurements ranging from 1 kGy to 1 MGy.

We have investigated the gamma radiation effects on parameters of electric-arc-induced long period fiber gratings in high Ge doped and B/Ge co-doped single mode fibers. The grating resonance wavelength shifts and amplitude of the dips of various cladding modes were monitored on-line to study the role of grating fabrication and fiber chemical composition. These studies lead to identification of boron as a critical core dopant for high radiation sensitivity. After a Co-60 gamma dose of 1 MGy, the optimized gratings show radiation-induced changes of their transmission dip wavelength up to 20 nm which is comparable to CO_2 laser-induced gratings reported by us previously [39]. These gratings also show very high temperature sensitivity specially when operated in dispersion turn-around-point (TAP) mode [38].

Fibers doped with different boron contents in SiO_2-GeO_2-B_2O_3 host were fabricated indigenously under collaboration with CGCRI, Kolkata, India. The gratings in such fibers were modeled and analyzed. We have also designed and fabricated a stable and robust sensor package unit for remote gamma dose measurements up to a dose of 1 MGy. Lab trials of such units have been carried out, and the experience in using such devices for dose estimation is discussed. These devices make arc-induced LPGs and CO_2 laser-induced LPGs in Boron doped fibers a strong candidate for applications in super Large Hadron Collider (LHC) and International Thermonuclear Experimental Reactor (ITER). **Table 2** shows the experimental results of such online measurements.

Figure 13 shows our on-line gamma dose effect measurements using specialty turn-around-point (TAP) long period fiber grating.

Wavelength before exposure (nm)	Wavelength measured after 4.30 h of dose (dose of 3.6 kGy) after removal from gamma chamber (nm)
1161	1165.4
1229.6	1231
1250	1253.6
1364	1366.4
1546.7	No significant dip

Table 2.
Gamma radiation exposure data for LPG of a 400 micron grating period inscribed in Fiber Logix, SM G652 fiber. Total dose: 65 kGy.

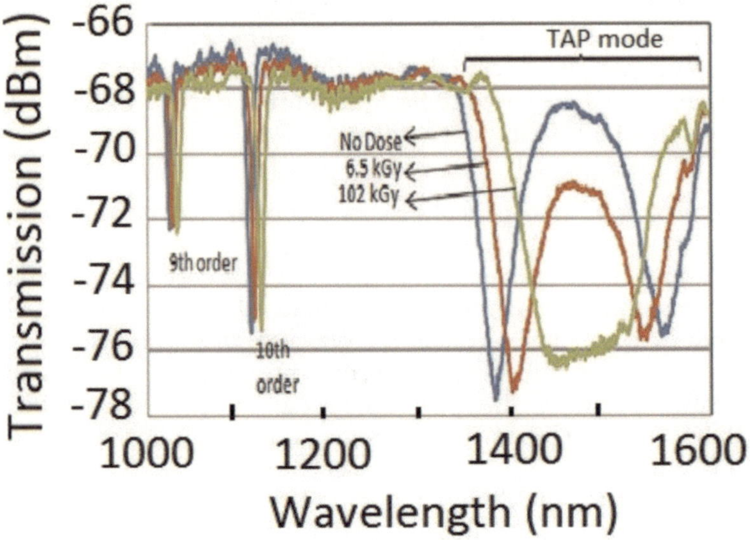

Figure 13.
Real-time resonance spectral measurements of TAP-LPG in gamma field [38].

The requirement of such sensors is divergent: for radiotherapy, we need very compact needle-like disposable tip sensors with very weak annealing properties. Once a patient is given a certain dose, the tip is discarded and a new dose sensing tip is used for other patients. For high dose sensing applications, the probe with high or low annealing may be desirable depending on application. For example, for long-term monitoring of integrated dose in waste storage facility will need very low annealing dose sensors while for repetitive dose measurements like gamma pencil fabrication, a high annealing rate sensor will be required.

6.6 Distributed gas sensors

The detection and identification of hazardous volatile compounds with low false alarm probability in large areas such as airports, underground metro stations, mines, bus stands, and major event arenas are a very daunting task. Other strategic applications include multi-point hydrogen leak detection systems for commercial and military launch vehicles that use cryogenic hydrogen as the main propellant. This gas is highly volatile, extremely flammable, and highly explosive. Hundreds of point gas sensors with electronic circuits are needed to cater big areas. Current detection systems use transfer tubes at a small number of locations through which gas samples are drawn and stream analyzed by a mass spectrometer. They are complex and costly systems and do not provide leak location.

Among several sensing systems, absorption-based systems are extensively explored and successfully employed. There have been serious efforts to develop absorption-based fiber optic systems for gas detection with higher sensitivity, fast response, and distributed sensing capability. They are based on measurement of VIS-IR-mid IR absorption by intended chemical species. The characteristic absorption spectra can be used as a fingerprint to identify a particular gas species. For example, many gas molecules have absorption lines in 0.8–1.8 μm band which is also a low loss transmission window of the silica fiber.

The spectral region of mid-IR is also a finger print region for most of the volatile compounds and nerve gases which is under investigation due to availability of compact tunable lasers in this band. The general principles of quantitative measurement of gas concentration have been described extensively in the literature [40, 41]. The basic principle is that the absorption is wavelength-dependent and compound specific. Initial systems used open path gas absorption cells comprising a pair of graded index collimator lenses with fiber pigtails [42]. The optical fibers here perform a passive role like transfer of light to and from absorption cell but play no active role in gas sensing. The National Aeronautics and Space Administration (NASA) reported a multi-point fiber optic multi-point hydrogen micro-sensor system for leak detection for launch vehicles [43]. The system consists of a multiplexed system having palladium coated micro lens attached to fiber tips. These chemically reactive coatings undergo changes in reflection in proportion to hydrogen concentration and are used as leak detectors. The invention of PCF or Holey fiber in 1990 opened new opportunities for exploiting the interaction of light with gases either through evanescent field or through hollow core region [44, 45]. The PCF family includes index guided PCF, hollow core photonic bandgap-fiber, and suspended core fiber.

6.6.1 Index guided photonic crystal fiber

Index-guided photonic crystal fibers are made with stack-and-draw process in fiber draw tower. They contain a periodic array of air holes running in cladding region and light is confined to solid core due to reduced effective index of the cladding. A fraction of evanescent field of power is extended into the holey region where it is absorbed by gas species and the gas concentration is obtained from intensity attenuation through the Beer-Lambert law. It is possible to design an IG-PCF that has a large fraction of evanescent power located within air-holes. This provides a good platform for gas sensing applications. Gas detectors based on absorption spectroscopy with IG-PCFs were experimentally demonstrated by Hoo et al. [46].

One concern in using IG-PCF as evanescent field sensors is the limited response time due to time required for sample gas to diffuse into holes from two open ends. Hoo et al. [46] have numerically calculated response time for acetylene gas in IG-PCF fiber (Λ = 1.55 μm, d = 1.4 μm) of length l with two ends open. When the fiber sensor length is 1 m, the time required to reach 90% concentration of exposed gas in surrounding is 200 min. This shows that we have to take a very short fiber if we require a response time of 1 min governed by diffusion time. However, this will reduce the sensitivity of the device due to limited path length. It is also observed that the relative sensitivity is also about 6% of that of open path cell per equal length. To achieve fast response time, one has to introduce periodic openings along the sensing fiber and measures like micro-pumping of sampling gases. To improve sensitivity, hollow core fibers may be a better choice as efficient interaction of guided light with gas molecules would be possible.

6.6.2 Hollow core photonic crystal fiber

Hollow core photonic crystal fiber is composed of a hollow core and a cladding with holes. The light is guided through the bandgap effect due to a proper design of hole sizes and gaps. They allow simultaneous confinement of optical mode and gas phase materials within the hollow core. This provides an excellent means for strong light/molecular interaction inside the fiber core over long distance.

6.6.3 Enabling technologies

Critical issues related to distributed, high sensitivity, fast response sensors are mainly governed by the time required for gas to diffuse into the holes and open ends of the sensor fiber. One can introduce a number of transverse holes by using micro-structuring techniques. Various techniques for transverse micro-structuring have been reported. Hensley et al. [45] have reported gas cell fabrication using femtosecond micro-machining. Tightly focused laser pulses are used to produce micro-meter diameter radial channels in hollow core photonic crystal fiber, and through these micro-channels, the core of the fiber is filled with test gas. There are some reports about fabrication of microchannel by using 193 nm ArF laser. Viola et al. [43] have reported micro-hole drilling in hollow core mid-IR transmitting fibers by means of high power CO_2 laser shots.

6.6.4 Recent developments and existing fiber-based systems

Distributed gas sensors rely on depositing specific gas sensitive coatings as cladding in multi-mode fibers or gratings and OTDR method. Recently, many designs and configurations using PCFs have been reported which targeted better sensitivity and distributed detection capability. The advantage of waveguide-based sensors over open path multiple reflection cells included low gas volume, high photon to molecular interaction, and reduction of fringe effects. Whitenett et al. [47] reported the operation of a 64 point fiber optic methane sensor installed on a landfill site in Glasgow, UK. Though the environmental conditions are harsh, the sensor has performed satisfactorily, detecting methane in the range of ~50 ppm to 100% methane. Viola et al. [42] have reported development of a distributed nerve gas sensor based on mid-IR spectroscopy. This has been possible due to availability of tunable quantum cascade lasers in the 9–10 μm range, suitable hollow fibers for this band, and low noise cooled detectors.

7. Conclusion

India is poised to use nuclear power in a big way. The safety of these power stations will depend on monitoring the radiation levels near plant and at waste disposal sites. In such environments, conventional sensors have certain limitations. Fiber-based sensors are being developed all over the world and are expected to make significant contribution to safe operation of nuclear fuel cycle. New technologies of laser micro- and nano-processing, mid-IR transmitting fibers, and hollow fibers have opened development of new structures and devices like fast response PCF-based gas sensors, high temperature Bragg gratings, distributed nerve gas sensors, and distributed nuclear radiation sensors for home land security. On the other hand, availability of all silica nuclear resistant fibers and nano-fibers and possibility of grating writing in such fibers using focused ion beam source and femto-sec lasers have attracted their use for different parameter monitorings like structural health monitoring in nuclear reactors, tokamaks, and storage facilities.

Acknowledgements

The authors are thankful for the help received from Shri Jai Kishore, Smt. Smita Chaubey, and Shri Sanjai Kumar of Fiber Sensors Lab., RRCAT, Indore, in conducting various experiments.

List of acronyms

AS	anti-Stokes
CGCRI	Central Glass & Ceramic Research Institute
DE	dual-ended
DNA	deoxyribonucleic acid
DWT	discrete wavelet transform
EMI	electromagnetic interference
ESM-PCF	endless-single-mode photonic crystal fiber
FBG	fiber Bragg gratings
FIR	finite impulse response
IIR	infinite impulse response
ITER	International Thermonuclear Experimental Reactor
LHC	Large Hadron Collider
LPGs	long period gratings
MOSFET	metal-oxide semiconductor field-effect transistor
NASA	National Aeronautics and Space Administration
OSA	optical spectrum analyzer
OTDR	optical time domain reflectometry
PCFs	photonic crystal fibers
RGs	regenerated gratings
ROFDTSs	Raman optical fiber-based distributed temperature sensors
RRCAT	Raja Ramanna Centre for Advanced Technology
SNR	signal-to-noise ratio
St	Stokes
TAP	turn-around-point
TDM	time division multiplexing
TLD	thermoluminescent dosimeter
UV	ultraviolet
WDM	wavelength division multiplexing

Author details

Sanjay Kher* and Manoj Kumar Saxena
Fiber Sensors Lab., FSOSS, Raja Ramanna Centre for Advanced Technology, Indore, India

*Address all correspondence to: kher@rrcat.gov.in

References

[1] Zhao XF, Li L, Ba Q, Ou JP. Scour monitoring system of subsea pipeline using distributed Brillouin optical sensors based on active thermometry. Optics and Laser Technology. 2012;**44**: 2125-2129

[2] Lopez RM, Spirina VV, Miridonova SV, Shlyagina MG, Beltran G, Kuzin EA. Fiber optic distributed sensor for hydrocarbon leak localization based on transmission/reflection measurement. Optics and Laser Technology. 2012;**34**: 465-469

[3] Culshaw B, Kersey A. Fiber-optic sensing: A historical perspective. Journal of Lightwave Technology. 2008;**26**: 1064-1078

[4] Grattan KTV, Sun T. Fiber optic sensor technology: An overview. Sensors and Actuators A Physical. 2000; **82**:40-61

[5] Dakin JP. Distributed optical fiber sensor systems. In: Culshaw B, Dakin J, editors. Optical Fiber Sensors: Systems and Applications. Vol. 2. Norwood, MA: Artech House; 1989. pp. 575-588

[6] Rogers AJ. Distributed optical-fiber sensors for the measurements of pressure, strain and temperature. Physics Reports. 1988;**169**:99-143

[7] Yilmaz G, Karlik SE. A distributed optical fiber sensor for temperature detection in power cables. Sensors and Actuators A Physical. 2006;**125**:148-155

[8] Hurtig E, Großwig S, Kühn K. Fiber optic temperature sensing: Application for subsurface and ground temperature measurements. Tectonophysics. 1996; **257**:101-109

[9] Höbel M, Ricka J, Wüthrich M, Binkert T. High-resolution distributed temperature sensing with the multiphoton-timing technique. Applied Optics. 1995;**34**(16):2955-2967

[10] Long DA. The Raman Effect. England: John Wiley; 2002. pp. 49-131

[11] Dakin JP, Pratt DJ, Bibby GW, Ross JN. Distributed optical fiber Raman temperature sensor using a semiconductor light source and detector. Electronics Letters. 1985; **21**(13):569-560

[12] Hill KO et al. Photosensitivity in optical fiber waveguides: Application to reflection fiber fabrication. Applied Physics Letters. 1978;**32**:647-649

[13] Kersey AD, Davis MA, Patrick HJ, LeBlanc M, Koo KP, Askins CG, et al. Fiber grating sensors. Journal of Lightwave Technology. 1997;**15**(8): 1442-1463

[14] Shimada Y, Nishimura A. Development of optical fiber Bragg grating sensors for structural health monitoring. Journal of Laser Micro/ Nano Engineering. 2013;**8**(1):110-114

[15] Kashyup R. Fiber Bragg Gratings. San Diego, CA, USA: Academic Press; 1999

[16] Kher S, Srikanth G, Chaube S, Chakraborty AL, Nathan TPS, Bhawalkar DD. Design, development and studies on Raman based fiber-optic distributed temperature sensor. Current Science. 2002;**83**(11):1365-1368

[17] Kher S, Srikanth G, Saxena MK, Nathan TPS. Development of distributed fiber optic temperature sensor with sub-meter resolution. Current Science. 2004;**86**(9):1202-1204

[18] Chakraborty AL, Sharma RK, Saxena MK, Kher S. Compensation for temperature dependence of Stokes signal and dynamic self-calibration of a

Raman distributed temperature sensor. Optics Communication. 2007;**274**: 396-402

[19] Chakraborty AL, Kher S, Chaubey S, Nathan TPS. Bidirectional frequency-domain digital filtering to simultaneously improve temperature resolution and eliminate spatial inaccuracy of a distributed temperature sensor. Optical Engineering. 2004; **43**(11):2724-2729

[20] Stoddart PR, Cadusch PJ, Pearce JB, Vukovic D, Nagrajah CR, Booth DJ. Fiber optic distributed temperature sensor with an integrated background correction function. Measurement Science and Technology. 2005;**16**(6): 1299-1304

[21] Saxena MK, Raju SDVSJ, Kher S, Arya R, Ravindranath SVG, Kher S, et al. Optical fiber distributed temperature sensor using short term Fourier transform based simplified signal processing of Raman signals. Measurement. 2014;**47**:345-355

[22] Guang-yong X, Jian-ping Y, Kun-jun Z, Wen M. Monitoring data processing of distributed optical fiber temperature sensor based on multi-wavelet. In: Proceedings of the International Workshop on Education Technology and Training & International Workshop on Geoscience and Remote Sensing. 2008. pp. 203-206

[23] Saxena MK, Raju SDVSJ, Arya R, Pachori RB, Ravindranath SVG, Kher S, et al. Raman optical fiber distributed temperature sensor using wavelet transform based simplified signal processing of Raman backscattered signals. Optics & Laser Technology. 2015;**65**:14-24

[24] www.google.co.in/search?q=BACKSCATTERED+LASER+LIGHT&source=lnms&tbm=isch&sa=X&ved=0ahUKEwiCvf-iq9jcAhWKT30KHdJ4CjYQ_AUICigB&biw=1093&bih=530

[25] Fukuzawa T, Shida H, Oishi K, Takeuchi N, Adachi S. Performance improvements in Raman distributed temperature sensor. Photonic Sensors. 2013;**3**(4):314-319

[26] Fernandez AF, Rodeghiero P, Brichard B, Berghmans F, Hartog AH, Hughes P, et al. Radiation-tolerant Raman distributed temperature monitoring system for large nuclear infrastructures. IEEE Transaction on Nuclear Science. 2005;**52**(6):2689-2694

[27] Lee CE. Self-calibrating technique enables long-distance temperature sensing. Laser Focus World. 2007;**43**: 101-117

[28] Suh K, Lee C. Auto-correction method for differential attenuation in a fiber-optic distributed-temperature sensor. Optics Letters. 2008;**33**(16): 1845-1847

[29] Zhang H, Wang Q, et al. Fiber Bragg grating sensor for strain sensing in low temperature superconducting magnet. IEEE Transactions on Applied Superconductivity. 2010;**20**(3): 1798-1801

[30] Marcelo M, Allil RCSB, de Nazar FVB. 14. Chapter "A Guide to Fiber Bragg Grating Sensors": In: Current Trends in Short- and Long-period Fiber Gratings. 2013

[31] Gusarov AI, Starodubov DS, et al. Comparative study of MGy dose level γ-radiation effect on FBGs written in different fibers. In: Proceedings of the OFS-13; April 12–16, 1999; Kyongju, Korea

[32] Gusarov AI, Chojetzkib C, et al. High total dose radiation effects on temperature sensing Fiber Bragg gratings. IEEE Photonics Technology Letters. 1999;**11**(9):1159-1961

[33] Krebber K et al. Fiber Bragg gratings as high dose radiation sensors.

Measurement Science & Technology. 2006;**17**:1095-1102

[34] Kher S, Chaube S, et al. Fiber optic temperature-insensitive, strain sensors for nuclear applications. International Journal of Applied Engineering and Technology. 2011;**1**(1):61-67. Available from: http://www.cibtech.org/J-ENGINEERING-TECHNOLOGY/PUBLICATIONS/2011/Vol%201%20No.%201/16-03-JET-Kher.pdf

[35] Henschel H et al. High radiation sensitivity of Chiral long period gratings. IEEE Transactions on Nuclear Science. 2010;**57**(5):2915-2922

[36] Rego G et al. Effect of ionizing radiation on the properties arc-induced long period fiber gratings. Applied Optics. 2005;**44**(29):6258-6263

[37] Gusarov AI et al. Comparative study of MGy dose level γ-radiation effect on FBGs written in different fibers. In: Proceedings of the OFS-13; April 12–16, 1999; Kyongju, Korea

[38] Kher S, Chaubey S, Kashyap R, Oak SM. Turnaround-point long period fiber gratings (TAP-LPGs) as high radiation dose sensors. IEEE Photonics Technology Letters. 2012;**24**(9):742-744

[39] Kher S, Chaubey S, Oak SM. Long period fiber grating based nuclear radiation sensors for high level dose applications. Journal of Instrumentation Science & Technology. 2013;**41**(2):135-142

[40] Bernath FP. Spectra of Atoms and Molecules. New York: Oxford University Press; 2005

[41] Svanburg S. Atomic and Molecular Spectroscopy: Basic Aspects & Practical Applications. Berlin: Springer; 2004

[42] Ho HL, Jim W, Demoken MS. Quantitative measurement of acetylene by using external cavity tunable diode laser. In: Proc. SPIE 3852, Harsh Environment Sensors II; Boston, US. 1999. pp. 124-133

[43] Viola R, Liberatore N, et al. Distributed nerve gases sensor based on IR absorption in hollow optical fiber. SPIE Proceedings. 2010;**7838**:78380H

[44] Kazemi AA, Goepp JW, et al. Fiber optic microsensro hydrogen leak detection system on aerospike X-33. In: SPIE Proceedings 6758, Photonics in the Transportation Industry: Auto to Aerospace; Oct. 2007

[45] Hensley CJ, Broaddus DH, et al. Photonic bandgap fiber gas cell fabricated using fs micromachining. Optics Express. 2007;**15**(11):6690-6695

[46] Hoo YL, Jin W, et al. Design and modeling of a photonic crystal fiber gas sensor. Applied Optics. 2003;**22**(18):3509-3515

[47] Whitenett G et al. Optical fibre instrumentation for environmental monitoring applications. IOP Journal of Optics A: Pure and Applied Optics. 2003;**5**:S140-S145

Chapter 5

Real-Time Particle Radiography by Means of Scintillating Fibers Tracker and Residual Range Detectors

Domenico Lo Presti, Giuseppe Gallo, Danilo Luigi Bonanno, Daniele Giuseppe Bongiovanni, Fabio Longhitano and Santo Reito

Abstract

In this chapter, a detailed description of the construction and the procedure for the measurement of performances of a charged particle imaging system is given. Such a system can be realized by the combined use of a position sensitive detector and a residual range detector. The position sensitive detector is made up of two superimposed and right-angled planes, each of which subsists of two layers of pre-aligned and juxtaposed scintillating fibers. The selected 500 μm square section fibers are optically coupled to two silicon photomultiplier arrays adopting a channel reduction system patented by the Istituto Nazionale di Fisica Nucleare. The residual range detector consists of 60 parallel layers of the same fibers used in the position detector, each of which is optically coupled to a channel of silicon photomultiplier array by means of two wavelength-shifting fibers. The sensitive area of both detectors is 90×90 mm^2. The performance of the prototypes was tested in different facilities with protons and carbon ions at energy up to about 250 MeV and rate up to about 10^9 particles per second. The comparison between simulations and measurements confirms the validity of this system. Based on the results, a future development is a real-time radiography system exploiting high-intensity pencil beams and real-time treatment plan verification.

Keywords: real-time particle radiography, innovative detector readout strategy, beam monitoring, scintillating optical fibers, treatment plan verification

1. Introduction

Hadron therapy is a promising alternative in treatment of tumors, because it is one of the most effective techniques of external radiation therapy, which allows killing tumor cells while leaving almost intact the surrounding issue. In order to achieve the maximum effectiveness, high precision is needed in dose delivery, which requires a real-time adequate quality control of the beam parameters (position, profile, fluence, energy) combined with the precise measurement of patient positioning [1, 2]. QBeRT is a particle tracking system [3] and consists of a position-sensitive detector (PSD) and a residual range detector (RRD) (see **Figure 1**). The main parts of this system are detectors expressly designed to achieve high-resolution imaging, high-resolution residual range measurement, large sensitive area, and high-rate beam compliance. The QBeRT system performs all these tasks and, advantageously, requires a low number of readout channels, making possible the reduction of the complexity and cost of the electronic data acquisition (DAQ) chain, by means of a readout channel reduction system patented by Istituto Nazionale di Fisica Nucleare (INFN) [4]. Both detectors, PSD and RRD, are able to work in imaging conditions, with particle rate up to 10^6 particles per second, and in therapy conditions (up to 10^9 particles per second). In imaging condition, the system is capable to realize a particle radiography and permits a real-time monitoring of the patient position in treatment room. In therapy condition, the PSD acts as a profilometer, detecting the position, the profiles, and the fluence of the beam. The combined use of the information measured by the PSD and the RRD allows to check the treatment plan in real time. The design of both detectors is based on scintillating optical fibers (SciFi) with 500 μm nominal square section.

The working principle of the scintillating optical fibers is schematized in **Figure 2**. SciFi consist of a polystyrene-based core and a PMMA cladding. The scintillating core is a mix of polystyrene and fluorescent dopants selected to produce the scintillation light when a particle releases energy in it and sets the optical characteristics for light propagation in the fiber. Scintillation light is produced isotropically but only a portion of these photons, in the two opposite directions along the fiber, can propagate by total internal reflection (TIR) mechanism. Multi-clad fibers have a second layer of cladding that has an even lower refractive index and permits TIR at a second boundary. External EMA is an optional external layer used to eliminate optical cross talk. SciFi sizes range from 0.25 to 5 mm square or round cross-sections and available in canes, spools, ribbons, and arrays.

The scintillation light is routed by the SciFi in the PSD, by means of wavelength-shifting fibers in the RRD, toward two silicon photomultiplier (SiPM) arrays, which output a proportional electric signal. PSD and RRD employ a DAQ chain divided in two sections. The first section consists of the front-end (FE) boards, which process the electric signal from the light sensor and perform the analog-to-digital conversion.

Data from the FE is acquired by a readout (RO) board based on a National Instrument system on module (SoM) for pre-analysis and filtering. The actual readout channel reduction scheme applied to the PSD limits the performances of

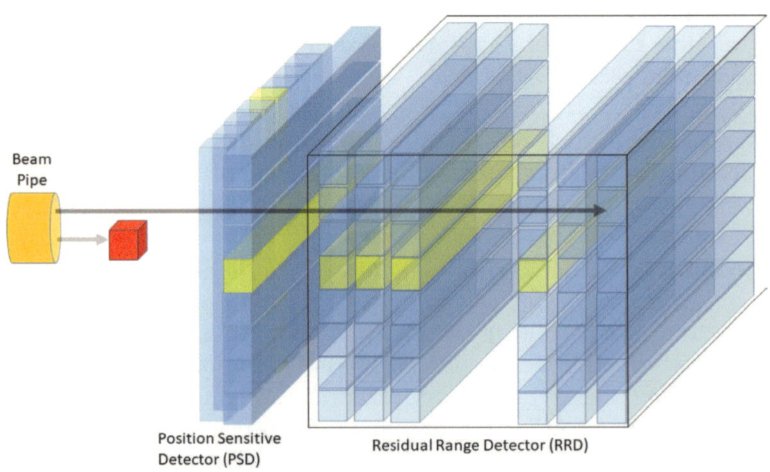

Figure 1.
Schematic of the QBeRT proton tracking system.

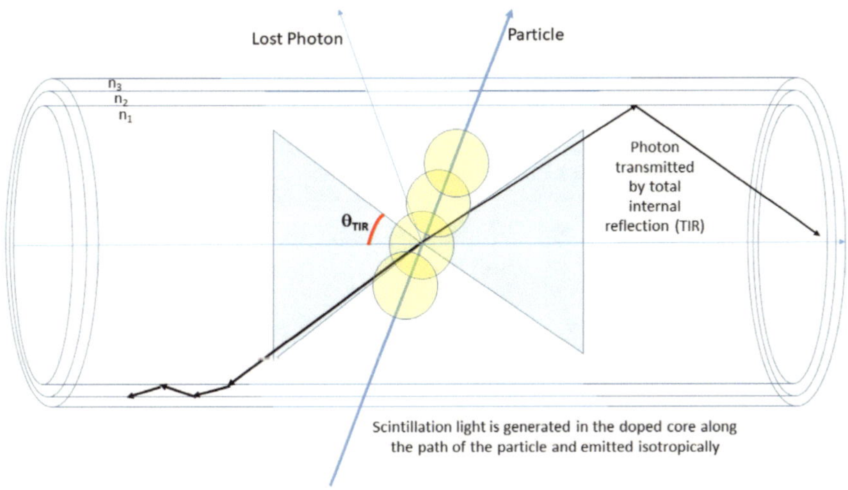

Figure 2.
Working principle of the scintillating fibers.

the detector when the beam spot size exceeds about 2 cm. Notice that the choice of the suitable readout channel reduction scheme and the modularity of the detector architecture allows to resize the sensitive area and the maximum beam spot size in order to fit any specific requirement. It is possible to obtain a large area detector (up to 400 × 400 mm^2) covering a range up to 250 MeV protons with high spatial and range resolution (up to 150 and 170 μm, respectively). The detectors described in this chapter have been intensively tested in imaging and therapy (up to 10^9 proton/s) conditions [5] thanks to the collaboration with the colleagues working at CATANA facility (Centro di AdroTerapia ed Applicazioni Nucleari Avanzate at Laboratori Nazionali del Sud-Istituto Nazionale di Fisica Nucleare in Catania), at TIFPA (Trento Institute for Fundamentals Physics Applications) in Trieste, and at CNAO (Centro Nazionale di Adroterapia Oncologica) in Pavia.

2. The position-sensitive detector

The PSD prototype has a sensitive area of 90 × 90 mm^2, which is made of two ribbons, layers of pre-aligned BCF-12 SCIFI, manufactured by Saint-Gobain Crystals, juxtaposed and orthogonally oriented, named the X and Y planes. The SciFi have 500 μm nominal square section. In detail, each single layer is composed of four ribbons of 40 fibers. The ribbons are optically isolated from each other by means of 220-μm-thick black adhesive tape to reduce cross talk between adjacent and overlapped ribbons. Each fiber is coated with white extra mural absorber (EMA) [6] to further reduce the cross talk between individual fibers. Particles intersecting the PSD's sensitive area deposit energy in the fibers which is partially converted in scintillation light. A fraction of this light is channeled in the core and propagated in the fiber toward the photo-sensor. When a particle loses suitable energy in all four SciFi layers, the coordinates of the intersection of its trajectory and the sensitive area can be measured. A picture of PSD detector is shown in **Figure 3**.

The PSD has 640 optical channels (four layers of 160 fibers each). The channel reduction system reduces the number of the readout channels without any data loss or degradation in the position measurement. The readout is performed in time coincidence, strongly reducing the effect of noise and chance coincidences, enhancing at the same time the performances of the system. The working principle of channel

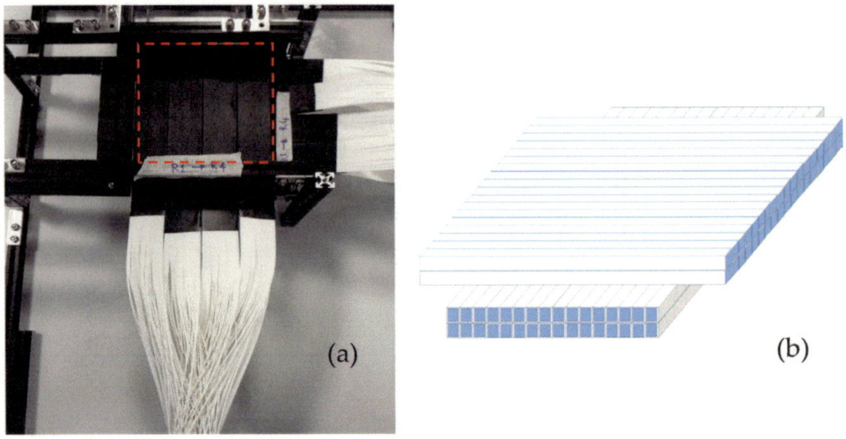

Figure 3.
(a) Picture of PSD detector during the assembly phase. The dashed red box highlights the sensitive area of the detector. (b) A sketch of the arrangement of the four layers is shown.

reduction can be argued by considering a strip detector able to detect one particle at a time. Each strip can be read from both ends, and the signals are grouped following the scheme reported in **Figure 4**, where a two-dimension 16-strip detector is illustrated. All strips are read out, on one end, in m groups of n neighboring strips, named *NeigSet*, while at the other end, the first strips of each group are grouped in *StripSet1*, the second strips of each group in *StripSet2*, and so on to m. The two numbers are not necessarily the same. The minimum number of total channels, obtained by choosing $n = m$, is $4\sqrt{N}$, where N is the total number of strips per layer, X or Y. This implies that the number of readout channels on the second side is equal to the number of fibers per channel on the first side. A particle crossing one

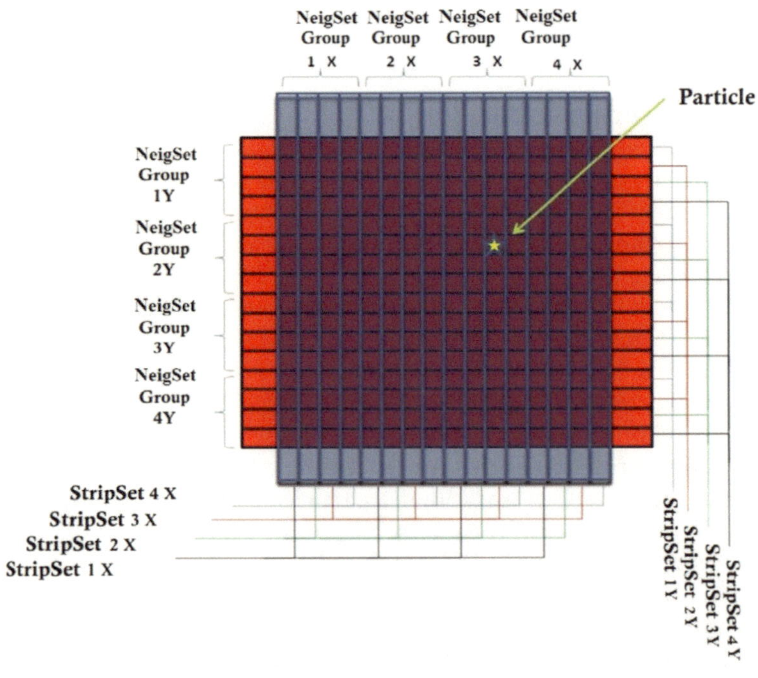

Figure 4.
An example of the application of the channel reduction scheme to a two-dimension 16-strip detector.

strip generates a signal at both ends of the fiber. Then, we have a signal from the ith *NeigSet* group and another from the jth *StripSet* that univocally identifies the $Strip_{hit}$ hit, according to equation (1):

$$Strip_{hit} = (i - 1) \cdot n + j. \tag{1}$$

The developed readout scheme reduces the number of readout channels to 16, instead of 32 channels without the application of the compression. When N is large, the reduction factor becomes important, allowing a compact and low-cost real-time acquisition. Notice that to reconstruct the point where the particle crosses the detector (event), an energy release in both planes is needed. In the PSD, readout occurs in time coincidence between the two layers of fibers for each plane. The channel reduction system can read the signal from the whole detector with only 112 channels (less than a fifth respect to the 640 total channels). Each of the 56 optical channels per plane is optically coupled to one of the 64 channels of a SiPM array. For a detailed description of the readout channel reduction system, see Refs. [7–10].

2.1 The PSD DAQ chain

As previously stated, the signals from the SiPM arrays are acquired by a DAQ chain divided in two main sections. The first section consists of the FE electronics which operate the analog-to-digital conversion. The digital-encoded data output from the FE is sent to the RO board which hosts a system on module (SoM) manufactured by the National Instrument (NI) for decoding and filtering. The SoM basically consists of a field-programmable gate array (FPGA) and a real-time processor, communicating by means of a direct memory access (DMA) data bus. The real-time processor has gigabit Ethernet connection for data transfer toward a PC, where the real-time visualization and storage of the results are accomplished.

Two FE boards, one for each direction, are required. Each FE board hosts a 64-channel SiPM array manufactured by Hamamatsu Photonics, mod. S13361-3050AE-08, with 3 × 3 mm^2 photosensitive area per channel. These custom designed boards amplify and filter the analog signals from each SiPM array channel and compare them, by means of an array of fast comparators, to an individual threshold, remotely settable by a DAC. An individual threshold per channel is useful to compensate the unavoidable mismatch of the SiPMs gain and of the optical coupling of SiPM to fibers. The output of each FE board is an asynchronous digital data bus of 56 bits which represents the status of the SiPM signals and is acquired by the NI SoM on the RO board. The FPGA samples the FE output at high frequency, up to 250 MHz. This data is transferred toward the SoM processor via DMA. The processor applies real-time filtering algorithms and, after discarding the spurious events, reconstructs the impact point of the particles. The SoM's FPGA can be programmed via a graphical approach by means of the LabVIEW platform. The LabVIEW platform also manages the entire acquisition chain and data processing in real time.

3. The residual range detector

The RRD prototype is a stack of 60 layers, 90 × 90 mm^2 area. Any layer is a ribbon of 180 BCF-12 SciFi. A picture of the detector is shown in **Figure 5(a)**. The ribbons are oriented horizontally and optically coupled at both ends to 1 mm square section wavelength-shifting fibers (WLS), see **Figure 5(b)** and **(c)**. To avoid optical cross talk between adjacent RRD layers, which would degrade range resolution and therefore energy measurement, each layer is optically isolated from the others by

Figure 5.
(a) View of the RRD prototype. (b) Detail of the optical coupling between SciFi and WLS fibers in a RRD layer. (c) View of the sensitive area of the RRD.

means of 100 μm black adhesive film. The SciFi used in the RRD are not coated with EMA because cross talk between adjacent fibers in the same layer does not affect detector resolution. The scintillation light produced in each layer at the passage of the particle is partially absorbed, reemitted, and channeled by the WLS. The two WLS coupled to a layer transfer the collected light to a channel of a SiPM array. The FE output is, then, processed by the DAQ chain. The SiPM array and the DAQ chain are identical to those used in the PSD described in the previous section.

A charged particle crossing the RRD passes through a number of layers as a function of its initial energy, before stopping. The dose deposited in each layer increases with depth up to the Bragg peak, where the particles produce the maximum amount of scintillation light. This point corresponds approximately to the end of the particles' path in the detector, so, by detecting the layer in which the light signal is the more intense, it is possible to measure their range. The RRD's working principle is reported in **Figure 6**. A calibration of the detector allows to obtain a

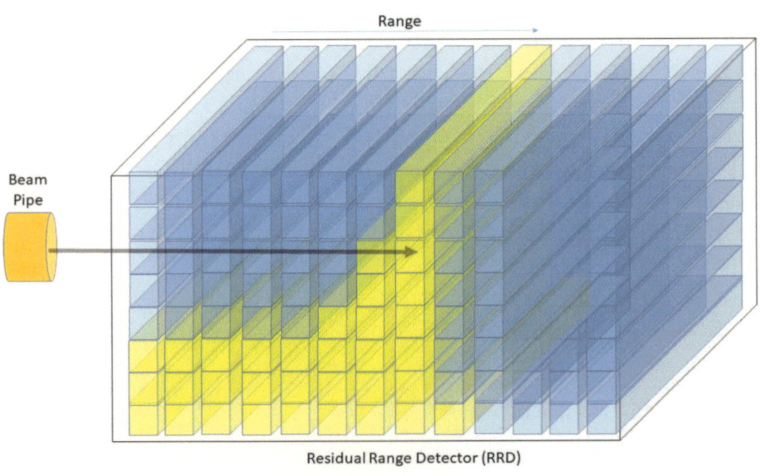

Figure 6.
Working principle of the RRD. In this figure, the height of the yellow column is proportional to the energy released in each layer by a particle crossing the RRD.

range-energy characteristic curve, which can be fitted by a function $R_0 = \alpha E^p{}_0$. In this way, it is possible to retrieve the initial energy of the particles from the measured range. For the actual RRD, the maximum measurable range was about 36 mm in polystyrene/PVC corresponding to the range of protons with 67 MeV initial energy, but this maximum range can be easily extended up to higher energies by placing a stack of calibrated water-equivalent range shifters between the beam exit and the RRD entry window.

4. Experimental results

Measurements at CATANA were carried out to fully characterize the performance of the prototypes, with protons up to 58 MeV at the output in treatment room. During the last year, other measurements have taken place at TIFPA proton irradiation facility. The spatial resolution of the PSD was measured by means of a calibrated brass collimator applied at the beam pipe exit in the treatment room at CATANA. After data analysis, it is possible to estimate the holes' centers and compare them with the projection of the collimator holes on the detector plane. Then, the mean distances between the reconstructed centers and the collimator hole centers were calculated for each hole, and the mean distance was about 130 μm, comparable with the (a priori) spatial resolution of the PSD, given by 500 $\mu m/\sqrt{12}$. The maximum spatial resolution is an intrinsic characteristic of the detector, independent of the readout strategy.

In order to calibrate the RRD, several measures of the range have been acquired changing the initial energy of protons. At CATANA facility, the proton beam energy can be passively modulated by placing a different calibrated PMMA range shifter at the beam pipe exit. The energy of the protons at the beam pipe exit was calculated by means of Monte Carlo simulation. The Bragg peak position is not exactly at the real end of the particles' path but just before. It is an experimentally consolidated practice to assume that the particle range measurement corresponds to the distance

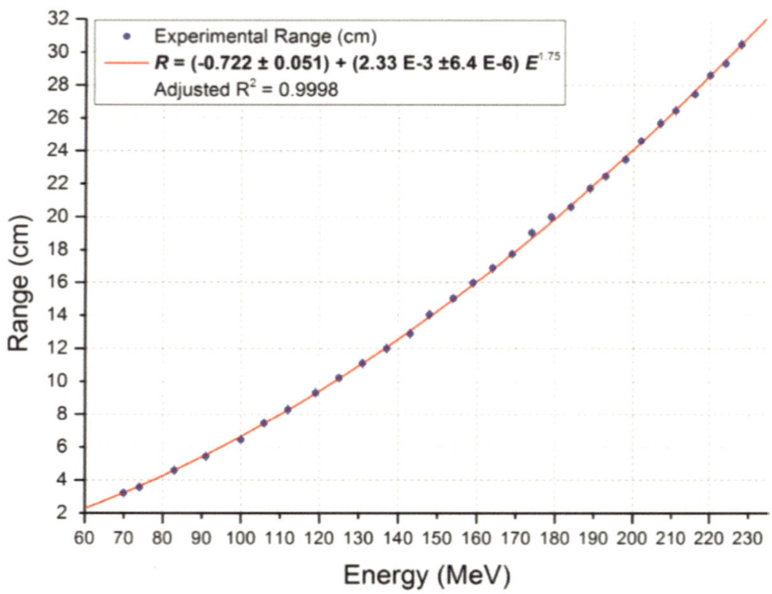

Figure 7.
Measured values of the range vs. the corresponding proton energy at TIFPA and resulting data fit.

from the entrance window where the intensity of the signal is one tenth of its maximum value.

This distance corresponds to the layer on the right of the Bragg peak (or the next layer compared to the incident beam direction). The results of this measurement are compared with the range values calculated by means of a Monte Carlo simulation of the response of the detector. Both data sets were fitted with the power law (Eq. (1)):

$$R = a + b \cdot E^{1.75} \tag{1}$$

where R is the range of the protons in the RRD, E is the kinetic energy at the entrance of the RRD, and a and b are free parameters of fit.

The same measurement was performed at TIFPA with proton energy in the range between 70 and 250 MeV and with a high-intensity beam, up to 10^9 protons per second.

In order to extend the range of the RRD, a series of calibrated water-equivalent range shifters, 10-mm-thick polystyrene slab phantoms was placed in front of the entry window of the RRD every time the energy of the beam exceeded the range of detector alone. In **Figure 7**, the range vs. energy calibration graph, measured at TIFPA, is shown.

5. Beam profile measurement

The PSD can work as a profilometer at rate up to 10^9 particles per second therapy conditions. It is able to measure the size and the position of the beam spot. As a

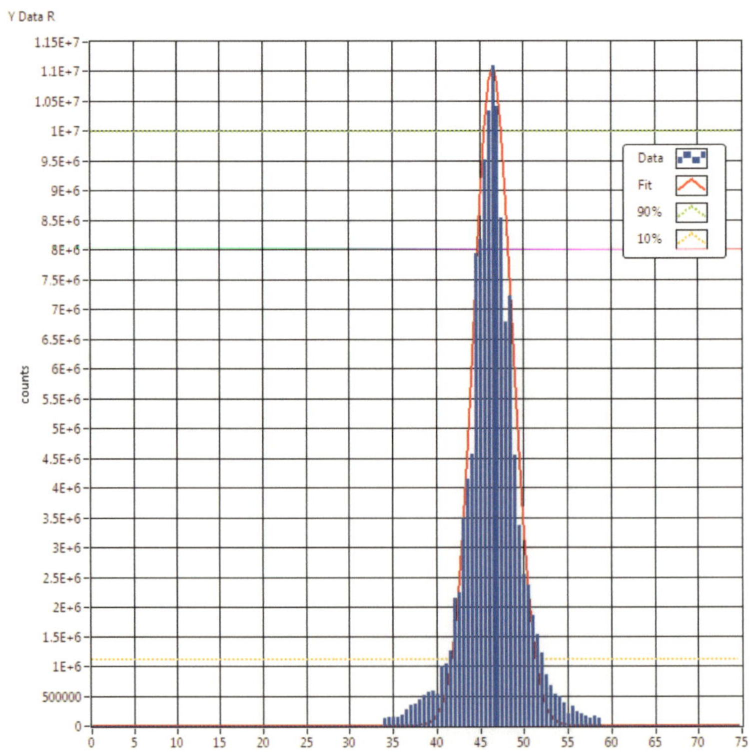

Figure 8.
Examples of Y profile of the proton beam spot at 70 MeV. The calculated Gaussian fit, in red, is superimposed to data.

consequence of the application of the channel reduction system, the beam profile can be reconstructed only if the beam spot size is lower or equal to the width of a ribbon (about 2 cm). The PSD was tested as a profilometer at TIFPA, where the beam optics causes a reduction of the beam spot size with increasing energy. In the test, the PSD worked properly at high energies, as shown in **Figure 8**.

6. The proton radiography

The PSD and RRD have been tested in radiography configuration at CATANA. In this test, the experimental setup is the one previously described, but the two detectors were simultaneously active.

In order to acquire a radiographic image, a range measurement must be performed for each particle crossing the PSD at a given position. Then, data acquisition must run at low beam intensity (imaging conditions, about 10^6 particles per second on average). When a particle causes a quadruple time coincidence in the PSD, the crossing position within the sensitive area is measured, and a trigger signal starts the measurement of the particle range in the RRD. The software analysis associates the positions measured by the PSD to the RRD range measurements event by event. At the end of data acquisition for each pixel, the software analysis calculates the centroid by Gaussian fit of the range measurements distribution corresponding to that pixel. The result of this analysis is, therefore, a 160 × 160 matrix, as many as the PSD pixels, in which each element is the centroid of the range measurement of the particles that have crossed the corresponding pixel. Note that the use of a single PSD placed before the RRD can introduce a not negligible error for the fact that the input and output particle crossing positions through the calibrated target one must necessarily be assumed coincident or

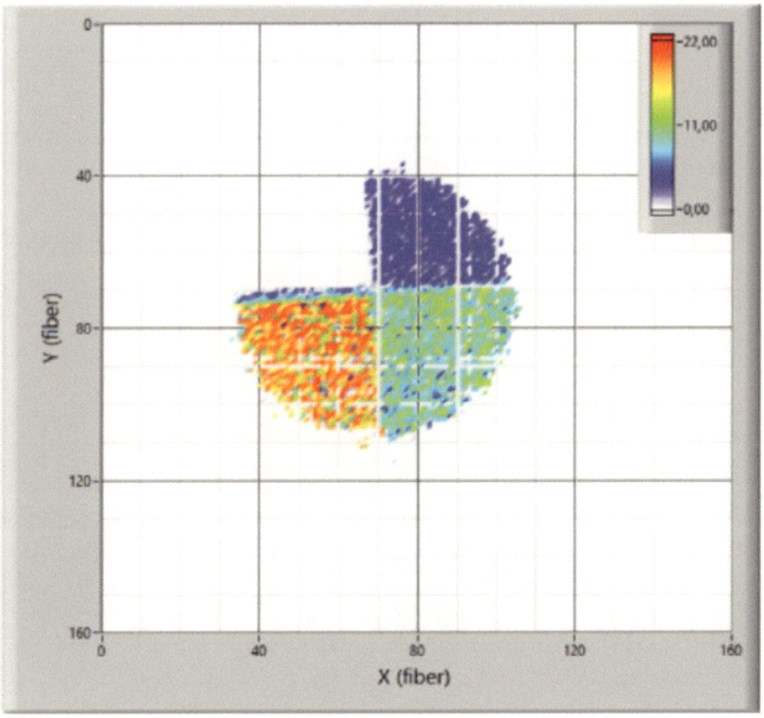

Figure 9.
The radiography of the ladder target with A12 range shifter.

undergone a negligible deflection traversing the medium. This error could be minimized using multiple PSD at different depths in the RRD.

A simple PVC target with the shape of a ladder was designed for the radiography test. Due to the homogeneous density of the target, in the radiography, only the differences in thickness traversed by the protons can be distinguished. The radiography image reported in **Figure 9** refers to a 3.5 cm diameter beam crossing a PMMA range shifter of about 10 mm thickness.

The z value in **Figure 9** is the centroid of the range distribution, expressed in numbers of RRD layers, pixel by pixel. Notice that the empty quarter-circle sector refers to the thickest step, 15 mm thick, of the ladder. The 58 MeV protons of the CATANA beam have insufficient energy to exit after passing through the thickness of the A12 range shifter and 15 mm of PVC. Moreover, border effects due to the non-orthogonality of the ladder with respect to the beam axis and the unavoidable divergence of the beam caused by the use of range shifters are visible in the radiography. The void pixels within the spot correspond to pixels where the range measurement statistics is too low. Many of these pixels are aligned along the same row or column, suggesting a correlation to low efficiency of the tracker in those areas. Two different 3D perspectives of the radiography are shown in **Figure 10(a)** and **(b)**. The last step in the analysis is the calculation of the relation between the measured range and the ΔE energy lost by the particles. The ΔE calculation must also take into account the energy lost by the particles in the PSD, which is placed between the target and the RRD. Since the sensitive areas of both detectors consist of 500 μm layers of SciFi, the PSD can be considered as an extension of the RRD. The residual proton range in the PSD and RRD was simulated as a function of the particle initial energy in the tracker E.

The range values thus obtained were fitted to the power law reported below in the equation, where R is the particle range in the RRD and PSD, expressed as the number of layers, and the resulting fit parameters are A = −0.191 ± 0.311 and B = 0.0370 ± 0.0006 (R - square = 0.998). Therefore, the energy loss ΔE can be easily calculated as

$$\Delta E[MeV] = 58 - \left(\frac{R - A}{B}\right)^{1/1.75}$$

The final radiography obtained after applying the energy-range conversion formula is shown in **Figure 11**.

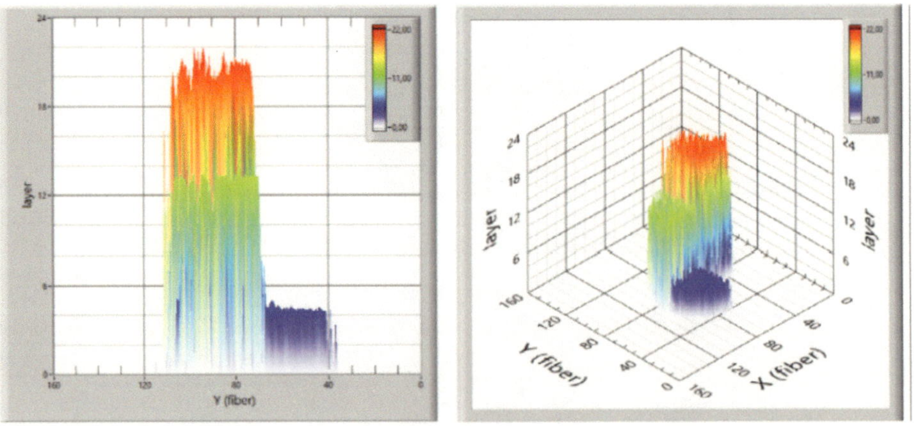

(a) (b)

Figure 10.
Two different perspectives of the 3D representation of the radiography: (a) lateral view and (b) isometric perspective.

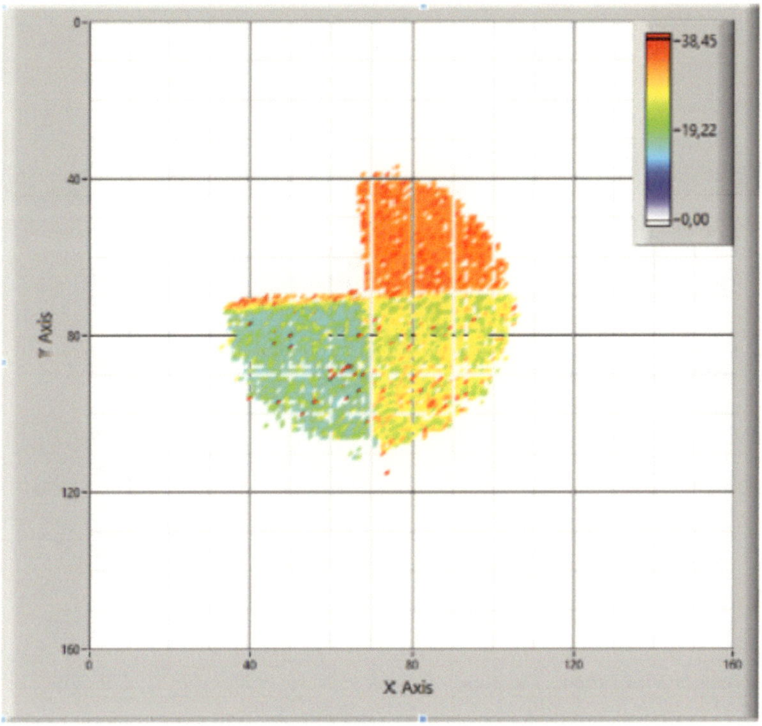

Figure 11.
The radiography of the ladder with A12 range shifter expressed in energy loss.

7. Radiograph data analysis

As mentioned earlier, radiography images reconstructed from range measurements are subject to some limitations: (i) lack of knowledge of the effective paths of the particles crossing the phantom because only one PSD was used. In this case, particle trajectories cannot be corrected according to the effect of Multiple Coulomb; (ii) further beam divergence was introduced by the tolerances in the alignment of the target, not exactly placed at isocenter and perpendicular to the incident beam direction. The reduction of the error in the calculus of the target thickness is obtained by the data filtering of range measurements. From the simulations, protons with an initial energy of 58 MeV crossing A12 range shifter, the target and the tracker, and stopping in the RRD have a maximum range straggling of σ_{str} = 0.4 mm, which already includes the effects of initial energy spread (0.3 MeV). So, in a region of interest (ROI) corresponding to a homogeneous quarter of the target, a range of measurements around the expected value from the simulation can be selected plus or minus two layers (equal to six times σ_{str}).

Subtracting the square of the maximum range straggling value of σ_{str} = 0.4 mm from the standard deviation of range measurements, it is again possible to find the a priori range resolution of about 170 μm. These mean range values can be converted into proton energy loss and subsequently into energy loss.

8. Future developments

The combined use of a pencil beam facility and the radiographic system, presented in this chapter, could allow the development of a faster real-time

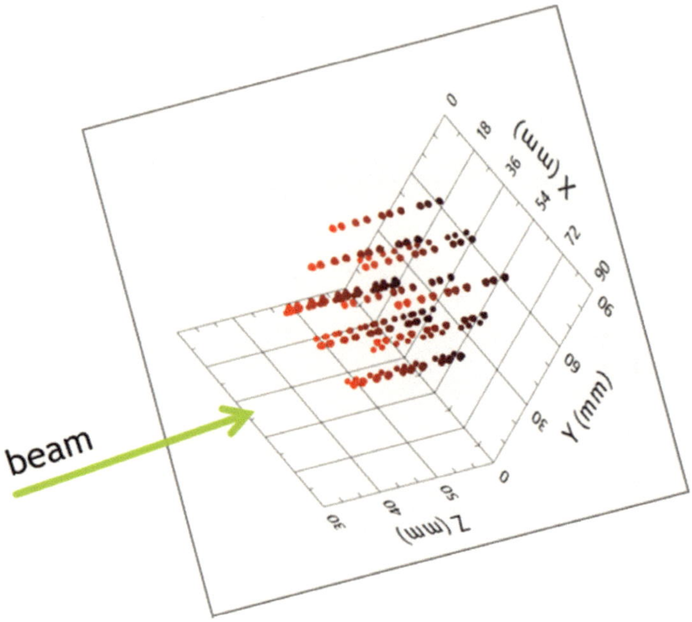

Figure 12.
The real-time reconstructed pattern. The x and y for each point are measured by the PSD. The z is the range measured by the RRD. The color is proportional to the measured fluence.

radiographic technique. Furthermore, the acquired radiography will be spatially correlated with the treatment plan applied to the patient. Exploiting the features of the described proton imaging system, a new method of quantifying treatment plan quality will be investigated.

A demonstrative measurement has been performed at CNAO in Pavia. A simple pattern of point in the field of view of the radiographic system, presented in this chapter, was covered by the pencil beam. The same pattern was modulated in energy, in the range of energy compatible with the range in the RRD, in order to obtain a 3D matrix. Each point in the matrix was covered by the pencil beam in one spill delivering a fixed dose, up to 10^9 protons per spill. The PSD measured the centroid, the FWHM and the fluence of the beam delivered in each position. The RRD measured the centroid, the FWHM of the range of the protons delivered in each spill. **Figure 12** shows the real-time reconstructed pattern.

The results demonstrate the potentiality of the system. Accurate measurements will be performed in order to refine these statements in a quantitative way at TIFPA in a treatment room. In these future tests, a calibrated phantom will be used for the measurement.

The definition of the optimal parameters for the radiography, e.g., beam energy and fluence to be chosen in order to obtain the required spatial and density resolution will allow the definition of the specifications for the design of the final detectors.

9. Conclusions

This chapter presents the design and characterization of an innovative imaging system for charged particle beam based on SciFi. The system consists of a position-sensitive detector and a residual range detector. Both prototypes, with a sensitive area of 90 × 90 mm^2, have cutting-edge performances, which distinguish them from all other devices designed for the purposes considered in this chapter. In addition, improvements in the DAQ chain and the use of SiPM arrays make possible the use of the PSD as a beam monitoring and quality assurance system, by measuring real time the center and the shape of the spot, the fluence, and residual energy of the beam. The verification of this feature was investigated and demonstrated in beam tests. The performance of the PSD and RRD was tested at CATANA proton therapy facility with energies up to 58 MeV. Moreover, Monte Carlo simulations of the RRD detector response and the radiography of a calibrated target were measured by the system. From the analysis of the results and by a comparison with data from simulations, the architecture and the technology were validated.

Tests at CNAO and TIFPA validated the functionality of these devices with active beam shaping systems using protons with energies up to 250 MeV. Future developments concern the real-time qualification of a treatment plan and the comparison of the results with those provided by the official dose delivery system. Furthermore, the feasibility of a real-time radiography exploiting pencil beam will be tested.

Author details

Domenico Lo Presti[1,2*], Giuseppe Gallo[1,2], Danilo Luigi Bonanno[2], Daniele Giuseppe Bongiovanni[2], Fabio Longhitano[2] and Santo Reito[2]

1 Department of Physics and Astronomy, University of Catania, Catania, Italy

2 Istituto Nazionale di Fisica Nucleare, Catania, Italy

*Address all correspondence to: domenico.lopresti@ct.infn.it

References

[1] Poludniowski G, Allinson N, Evans P. Proton radiography and tomography with application to proton therapy. The British Journal of Radiology. 2015;**88**(1053):20150134

[2] Schneider U, Pedroni E. Proton radiography as a tool for quality control in proton therapy. Medical Physics. 1995;**22**(4):353-363

[3] Gallo G, Lo Presti D, Bonanno D, Longhitano F, Bongiovanni D, Reito S, et al. QBeRT: An innovative instrument for qualification of particle beam in real-time. Journal of Instrumentation. 2016;**11**(11):C11014

[4] Lo Presti D. Detector based on scintillating optical fibers for charged particle tracking with application in the realization of a residual range detector employing a read-out channels reduction and compression method. INFN Patent No. WO2013186798; 2013

[5] Cirrone GAP et al. A 62-MeV proton beam for the treatment of ocular melanoma at Laboratory Nazionali del SUD–INFN. IEEE Transactions on Nuclear Science. Jun. 2004;**51**(3):860-865

[6] SGC Products. Scintillating Fiber Webpage. http://www.crystals.saint-gobain.com/products/scintillating-fiber

[7] Lo Presti D, Bonanno D, Longhitano F, Pugliatti C, Aiello S, Cirrone G, et al. A real-time, large area, high space resolution particle radiography system. Journal of Instrumentation. 2014;**9**(06):C06012

[8] Lo Presti D, Aiello S, Bonanno D, Cirrone GAP, Leonora E, Longhitano F, et al. OFFSET: Optical Fiber Folded Scintillating Extended Tracker. Nuclear Instruments and Methods in Physics Research Section A: Accelerators, Spectrometers, Detectors and Associated Equipment. 2014;**737**:195-202

[9] Lo Presti D, Bonanno D, Longhitano F, Bongiovanni D, Russo G, Leonora E, et al. Design and characterisation of a real time proton and carbon ion radiography system based on scintillating optical fibres. Physica Medica. 2016;**32**(9):1124-1134

[10] Lo Presti D et al. An innovative proton tracking system for qualification of particle beam in real-time. IEEE Transactions on Radiation and Plasma Medical Sciences. 2017;**1**(3):268-274. DOI: 10.1109/TRPMS.2017.2690842

Chapter 6

Review of Liquid-Filled Optical Fibre-Based Temperature Sensing

Fintan McGuinness, Gabriel Leen, Elfed Lewis, Gerard Dooly, Daniel Toal and Dinesh Babu Duraibabu

Abstract

While ever higher resolution temperature sensing is of demand in research and industry, the cost of sensors in the sub-milli-Kelvin (mK) range can be restrictive. Furthermore, as the majority of commercial temperature sensor measurements are transmitted via electrical circuits, significant post-processing is required to obtain a high-resolution due to phenomena such as electromagnetic interference, self-heating, electrical noise, etc. Consequentially, research in recent years has focused on the development of several technologies which overcome this issue, with optical fibre sensors proving to be a viable option. Owing to this, the following chapter will aim to review the current state-of-the-art in liquid filled optical fibre temperature sensing and the underlying methods.

Keywords: optical, fibre, temperature, sensor, liquid, review

1. Introduction

Prior to the development of thermometers, or indeed a scale of temperature, Renaissance Italy had already began to question the nature of relative temperatures with scientific postulations such as *Bardi's problem*. This problem was posed by Count Bardi di Vernio to Galileo to determine why a person felt cold upon entering a river in summer, yet grew comfortable over time. While forgoing a direct solution to Bardi's Problem, Galileo reportedly developed a thermoscope in response [1]. Although the device was capable of showing variation in sensible heat it differed from a thermometer in that it did not have a defined scale, it also suffered from barometric influences due to the nature of its construction [2]. It should be noted at this point that there remains uncertainty over the original inventor of the thermoscope, however four prime candidates have been identified, these being; Galileo, Sanctorius Sanctorius, Robert Fludd and Cornelius Drebbel [1]. The advent of precision thermometry originated with the designs of Fahrenheit in the early eighteenth century, with his sealed mercury-in-glass thermometers being a significant advancement on the then state-of-the-art [2]. While specialised liquid-in-glass thermometers have demonstrated resolutions in the region of 0.2 K [3] they are the least useful for accurate measurements in a temporal sense. This is primarily due to the inability to perform transient measurements and relatively cumbersome geometry when compared to electronic sensors.

Currently, industrial temperature sensor designs typically rely on either the thermoelectric effect or temperature dependant resistance. Owing to their wide operating range, thermocouples have been utilised in general engineering applications ranging from precision temperature control for tea processing factories to monitoring the cell temperatures of hepatoblastomas, with peak sensitivities in the region of 20 mK [4, 5]. However due to relatively poor response times, thermocouples are unsuited to high speed measurement. Another robust temperature sensor design is the negative temperature coefficient thermistor (NTC). A particular benefit to using NTCs is the ability to custom mould them for a desired application along with easily tunable resistance at the device and circuit levels [6]. Clow et al. [7] presented an array of NTCs used as a single sensor which was capable of 0.1 mK resolution with application in paleothermometry. While the sensor had excellent resolution it suffered from several drawbacks, the most relevant being a 7 s response time, which prevented it from recording highly transient temperatures gradients. Furthermore the sensor was sensitive to triboelectric effects from snow deposits on sensing wires and self-heating related noise, these being issues which scaled with distance. Precision resistance temperature detectors (RTDs) with milli-Kelvin (mK) resolution are readily available with a response time of 950 ms, and high-speed RTDs in the region of 400 ms [8]. While these sensors present high precision over their working range, a drawback of using RTDs is their inherent fragility. While suited for use in a laboratory, and other controlled environments, general deployment is still proving difficult at present. Hewlett & Packard pioneered the initial development of quartz thermometers in the mid-1960s, with resolution being in the range of 0.1 mK [9]. While capable of achieving sensitivity equivalent to the sensor presented by Clow et al. [7] the reported response time was 30% slower, and had a larger overall footprint.

The issues posed by the use of these *traditional* sensors in the presence of electromagnetic interference (EMI) and other considerations such as damage resilience and projected lifespan, has resulted in significant investment being made in the exploration of optical fibres as a means of high resolution temperature sensing. While still relatively fragile compared to thermocouples and glass bulb thermometers, optical fibre temperature sensors (OFTS) offer the possibility of a cost effective, high speed and precise sensor. The purpose of this review is to present the current state-of-the-art in a subset of optical fibre sensors that is to say, liquid-filled optical fibre temperature sensors (LiF-OFTS). Further to this, applications in which they have been utilised will be presented along with a discussion of potential future developments.

1.1 Temperature sensor requirements

As temperature sensors can be a mission critical piece of equipment various standards have been issued with respect to their usage. Common publishers of temperature sensor requirements for calibration and reporting include the ASTM [10] and IEEE [11]. The measurement range of a given sensor is largely determined by the nature of its operating environment. In an industrial process the range may be several hundred Kelvin, whereas a homoeostatic biomedical thermometer requires a range in the order of only 15 K [12]. Some commercially available probes made from exotic materials such as a tungsten-rhenium alloy thermocouples (Types G, C and D) have wide working ranges, approximately 273–2590 K [13]. These commonly find application in processes involving high temperatures, however require further protection if being utilised in an oxidising environment.

The overall resolution of a temperature sensor may be more important than the operating range depending on an intended application. While it is not critical that a domestic oven have precision temperature control, it can be of great assistance to

Company	General Electric	Hanna Instruments	SENSTECH
Sensor type	PRT [18]	TC (K-Type) [8, 19]	NTC [20]
Min. temperature	283 K	73 K	218 K
Max. temperature	450 K	573 K	398 K
Resolution	0.003 K	0.1 K	0.044 K
Response time	0.5 s	4 s	0.75 s
Accuracy	0.5 K	0.5 K	0.5 K

Table 1.
Commercially available high resolution temperature sensors.

pharmaceutical calorimetry. A review of the topic conducted by Buckton et al. [14] highlighted the need for identical and constant temperatures within the fermenter and calorimeter, along with maintaining an isothermal condition within the calorimeter itself. As the heat released in these reactions is typically of the order 1×10^{-4} mW it is readily evident that precision thermometry is beneficial in obtaining repeatable results. Rossi et al. [15] examined the relationship between intracranial and core temperatures in patients within 2–168 h post severe head injury. From this, intracranial temperatures were verified by comparison to the Delta OHM HD 9215, which had a resolution of 0.1 K. It was reported that core temperature was a poor indicator of intracranial temperature during pyrexic episodes, which was indicative of benefits to independent high resolution temperature sensing within the cranium. As magnetic resonance imaging (MRI) is common in patients post trauma, significant benefit would be provided should the thermometer be unaffected by the strong magnetic field, as it may remain in place to continue real-time monitoring.

Sampling rate is an important factor when deciding on a temperature sensor as it has to be considered in conjunction with the sensor response time. If the heat transfer from the system is known, then a rate of temperature change may be determined by Eq. (1) [16]. Evidently as temporal resolution is increased smaller variations in heat, and as a consequence temperature, may be determined.

$$\frac{\delta Q}{\delta t} = c \cdot \frac{\delta T}{\delta t} \tag{1}$$

Further to the items outlined above, specialised considerations may be required. In automotive engine development it is commonplace to embed multiple sensors inside the cylinder head/wall. While electronic sensors operate with little issue for the majority of a four-stroke cycle (720°) the spark plug activates for several degrees before or after top dead centre. As this is a period of strong pressure and thermal gradients, EMI introduced by the spark plug creates difficulty in analysing the recorded data [17]. A small sample of commercially available high resolution temperature sensors is presented in **Table 1**.

2. Advantages of optical fibre sensors

As mentioned in the introduction, optical fibre sensors present many advantages over their electronic counterparts. Therefore, the following section will examine several properties of optical fibres which indicate their viability as an alternative to electronic sensors.

2.1 Construction

Commercially available silica fibres are widely used in the development of optical fibre temperature sensors owing to their small cross-sectional area and consequentially their ability to be implemented in restricted areas. However, the relatively low cost of silica fibre compared to more exotic fibres such as those manufactured from fluoride or synthetic sapphire may be a factor to consider. As is reported, optical fibre sensors are particularly suited to environments where an electronic sensor may not have sufficient protection from EMI, and where line of sight to the measurement point is obstructed [21]. Furthermore, optical fibre sensors do not represent a potential source of ignition in an explosive environment; can be biocompatible; and can be made to work over very long distances. To date, optical fibre temperature sensors have been implemented in applications ranging from *in-vivo* biomedical sensing [22], to structural and geothermal engineering [23].

2.2 Measurement stability

In civil engineering applications, such as bridges, use of chirped fibre Bragg gratings (FBGs) provide excellent through life performance as fibre degradation is in excess of 25 years [24] with data transmission losses being minimal [23]. Furthermore, stability of measurements are generally quite good due to performance being driven predominantly by the wavelength stability of light source used. This being an easy factor to account for, as said light source may be kept in a controlled environment.

2.3 Distributed sensing

In addition to the EMI resistance and dimensional advantages of using optical fibre sensors, their ability to provide multipoint sensing with minimal use of fibre is a desirable characteristic. One such method is the use of chirped FBGs whereby multiple FBGs are inscribed on a single fibre, with the grating period modified at each location thus providing high spatial resolution. Wave division multiplexing (WDM) is another commonly used method in multi-point sensing with optical fibres, however this method narrows the usable bandwidth of light proportionally to the number of fibres used. Historically, Raman scattering has been used as an efficient means of multi-point temperature measuring; however, Brillouin sensors reported have shown exceptional strain measurement capability for an equivalent temperature sensing performance [25].

3. Theory and working principles

In the use of optical fibres for temperature sensing, measurements can be conducted using both intrinsic and extrinsic techniques. The most common technique for assessing temperature changes using optical fibres is the use of interferometry. Examples being fibre Bragg gratings (FBGs), extrinsic Fabry-Perot interferometers (EFPIs), Mach-Zehnder interferometers (MZIs) and Sagnac interferometers (SIs). As the fibre requires a light source to conduct the measurement, various types have been explored such as; broadband light source [26] monochrome laser [21] and swept laser [27]. Measurements can be carried out by observing either the transmitted spectrum, or the reflected spectrum. When recording the reflected spectrum losses occur in the back reflections through the fibre and at the coupler between the light source and spectrometer. That said however, this method allows for the fibre to be used as a point measurement device in applications where transmission is impractical, i.e. biomedical [28], automotive [29], and pharmaceutical [30] sensing. As the modus operandi of any particular optical fibre sensor is modulation of the light source, this can be carried out via intensity, frequency, or phase modulation. The latter of which will be the main focus of the review as this is the method typically employed by LiF-OFTS.

3.1 Fibre Bragg gratings

Fibre Bragg gratings are created by periodically modifying the refractive index of an optical fibre core. At each change of refractive index the reflected light constructively interferes producing a high intensity narrowband signal. This effect is described by Eq. (2) where λ_B is the Bragg wavelength, n_{eff} is the effective refractive index, and Λ is the pitch between each of the modified refractive indices. **Figure 1** provides a schematic of an FBG inscribed on a fibre core. As is evident from Eq. (2) care must be taken to eliminate, or account for, mechanical straining of the fibre as this will artificially modify the grating period. Once mechanical strain has been determined the change in Bragg wavelength with temperature is given by Eq. (3). A review on the packaging of FBGs is provided by Hong et al. [31].

$$\lambda_B = 2n_{eff} \cdot \Lambda \tag{2}$$

$$\frac{d\lambda_B}{dT} = 2\left(n_{eff} \cdot \frac{d\Lambda}{dT} + \Lambda \cdot \frac{dn_{eff}}{dT}\right) \tag{3}$$

Owing to the simple nature of their construction FBGs have been utilised to great success as a means of sensing temperature, however they are not without inherent issues such as damage due to exposure to excessive temperatures [22], and grating orientation to the heat source [32]. Zhang et al. [22] reported the assessment of cylinder head temperature and mixture flow within maritime diesel engines, where significant steps were taken to protect the fibre coating from excessive cylinder head temperatures (873.15 K). Gassino et al. [32] examined the use of FBGs in the presence of strong temperature gradients (~10 K/cm) during thermal ablation of tumours. Several key design factors were discussed from which it was determined the largest sources of error were caused by the temperature gradient along the length of the FBG, and FBG orientation with respect to the temperature gradient.

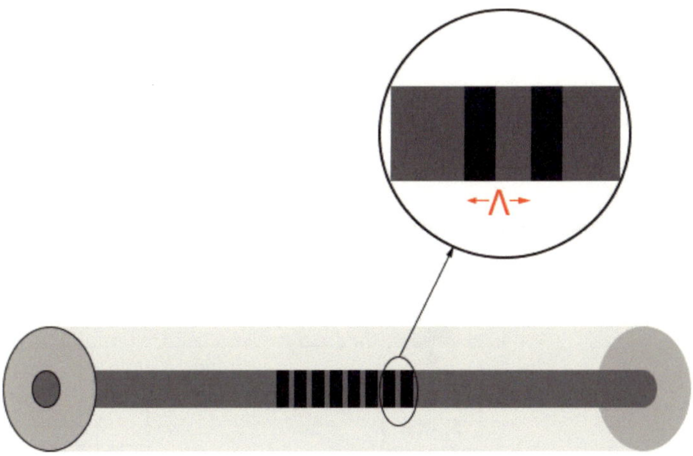

Figure 1.
Schematic of FBG, highlighting grating pitch (Λ) in fibre core.

3.2 Extrinsic Fabry-Perot interferometers

Fabry-Perot based sensors are typically extrinsic in nature and used as a point measurement device at the tip of an optical fibre. A common construction of EFPI sensors involves a single mode fibre (SMF) spliced to a capillary with a multi-mode fibre (MMF) which acts as a diaphragm fused to the opposing end of the capillary, creating an air filled cavity between the SMF and MMF. **Figure 2** highlights the dimensions relevant to diaphragm deflection and rigidity, with the construction presented in **Figure 3**. As light reaches the end face of the SMF a portion of the light is reflected, with the remainder transmitted into the air cavity. Similar reflections occur at the air-diaphragm, and diaphragm-external media interfaces. As the light reflected from the inner and outer faces of the diaphragm has travelled a greater distance, than the light reflected at the end face of the SMF, a phase difference between reflections will exist. While FPIs behave similar to FBGs in that they have varying refractive indices along the axial direction, adjustment of the diaphragm thickness can be used as a means of increasing pressure sensitivity by modifying the diaphragm flexural rigidity. Said flexural rigidity is determined by Eq. (4) [33] where h, E and ν are the diaphragm thickness, Young's modulus, and Poisson's ratio respectively. Eq. (5) [33] provides the relationship between diaphragm displacement and uniformly applied pressure, this being fundamental to the temperature sensor presented by Poeggel et al. [34]. From this it becomes apparent that a thinner diaphragm leads to greater maximum deflection, and said maximum deflection occurs at the diaphragm centre.

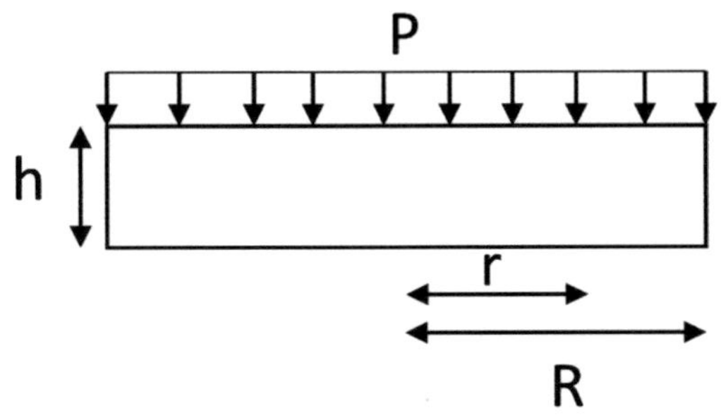

Figure 2.
Dimensions which influence diaphragm flexural rigidity and deflection.

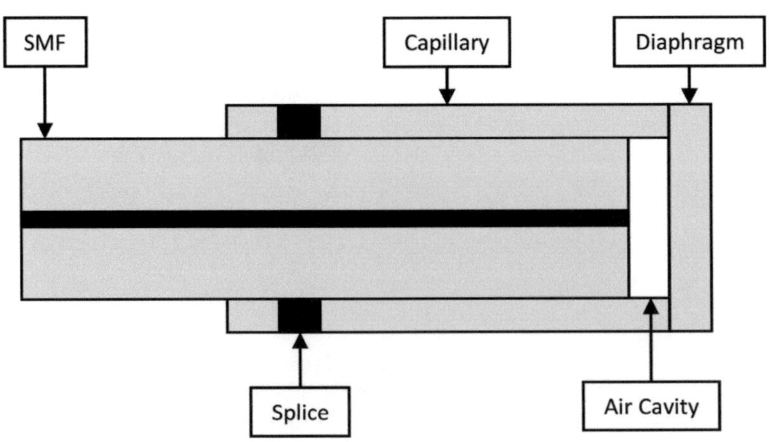

Figure 3.
Schematic of typical EFPI.

$$D = \frac{Eh^3}{12(1-\nu^2)} \tag{4}$$

$$z(r) = \frac{3}{16} \cdot \frac{(1-\nu^2)(R^2-r^2)}{Eh^3} \cdot P \tag{5}$$

Gao et al. proposed an alternative construction, using particle vapour deposition (PVD) to create a 'diaphragm', which was fused directly to an MMF [35]. This construction provides a number of advantages, such as a robust construction, and the ability to custom tune optical path lengths without the uncertainties related to splicing. Eq. (6) presents the relationship between the optical path differences (OPD) with the thermal coefficients of refractive index (α_n) and material thickness (α_d) for the PVD diaphragm.

$$\frac{OPD(T)}{OPD(T_0)} \approx 1 + (\alpha_n + \alpha_d)(T - T_0) \tag{6}$$

3.3 Mach-Zehnder interferometers

Mach-Zehnder interferometers (MZIs) operate by splitting the source light and introducing an optical path difference, before recombining the beams prior to the detector [36]. As a means of sensing in optical fibres, common practice has been to splice a length of photonic crystal fibre (PCF) between two lengths of SMF [37–40]. These operate by allowing the core mode of the SMF to be split into core and cladding modes at the first splice point within the PCF. These are subsequently rejoined at the second splice point with an interference between the two modes becoming apparent in the transmission, this is usually represented by resonant dip (s) in the signal transmitted to the optical spectrum analyser (OSA). Modification to the air holes of the PCF is commonly carried out via collapsing or filling with various fluids, in order to increase sensitivity to the desired measurand. For the purpose of temperature sensing it is common to fill the PCF holes with a fluid which has a high magnitude (as it may be positive or negative) thermo-optic coefficient [37, 39, 41, 42]. **Figure 4** below provides a schematic of how a typical MZI sensor setup is presented in literature. In a PCF the two branches of the MZI are the cladding and core modes respectively. A transverse section of a PCF is presented in **Figure 5** where the core, and cladding air holes are highlighted.

3.4 Sagnac interferometers

Sagnac interferometers (SIs) behave similarly to MZIs in that they compare optical path differences of two beams which have been split and subsequently recombined. However, they differ such as the two beams counter-propagate with

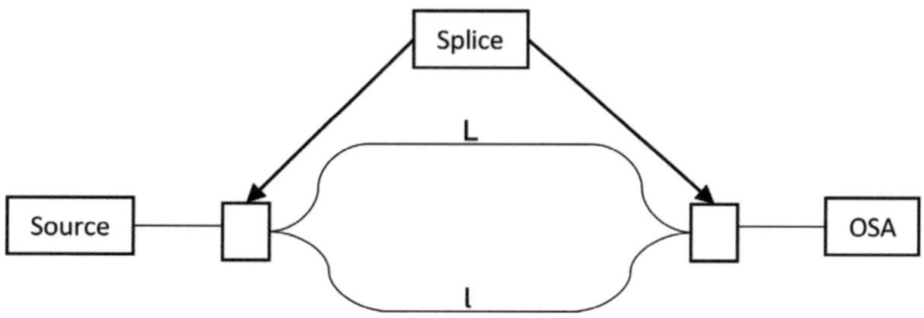

Figure 4.
Schematic of Mach-Zehnder based sensor.

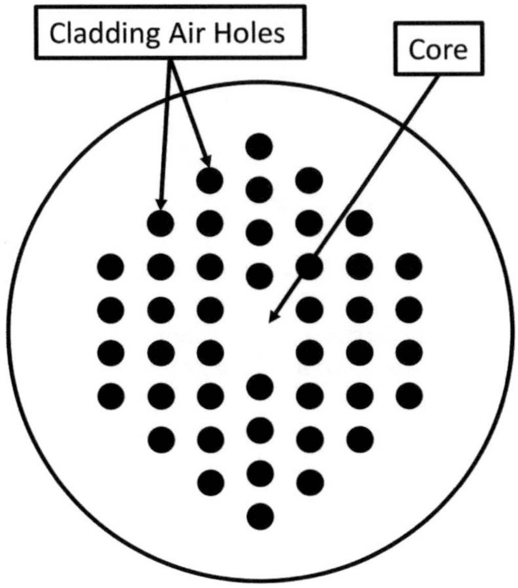

Figure 5.
Transverse section of a photonic crystal fibre.

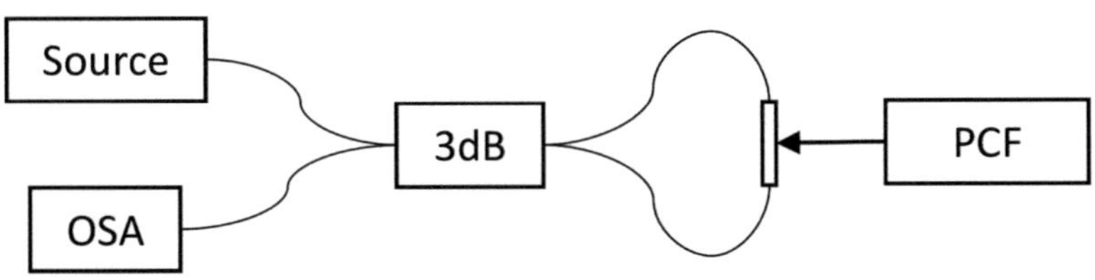

Figure 6.
Schematic of potential Sagnac interferometer based sensor.

respect to each other before being recombined at the detector. This type of interferometer is the basis of fibre-optic gyroscopes [43]. While it has been utilised in thermometry [44, 45] the sensors reported have not seen a similar level of research compared to their MZI counterparts. Reported sensors used a source beam counter-propagated by a 3 coupler. Birefringence in the PCF resulted in the optical path difference which was detected by the OSA [44, 45]. **Figure 6** above highlights how an SI based sensor may be constructed using optical fibre equipment.

4. Liquid filled optical fibre temperature sensors

As evident from Sections 1 and 3 temperature sensors have a long history. With respect to LiF-OFTSs however, while several types of sensor and base technologies have been presented in the literature, the authors have not encountered a concise review of said sensors. The materials utilised have included; alcohol [44] motor oil [34], silicone (polydimethylsiloxane) [46], and immersion oils [39]. While polydimethylsiloxane (PDMS) has only been presented in its cured form, it is not inconceivable that the liquid form be used given its optically transparent nature. The following section is arranged by the base interferometry principle of each fluid filled sensor.

4.1 EFPI type sensors

Chen et al. presented an FPI construction which contained an air micro-bubble encased in cured PDMS [46]. The sensor was manufactured by splicing an SMF to a hollow core fibre (HCF) 282 μm in length. The PDMS was subsequently introduced into the HCF via capillary effect. As PDMS entered the HCF it formed the air micro-bubble with the SMF, where the length of air micro-bubble was controlled by the period of time PDMS was allowed to enter the HCF. Online monitoring was conducted to establish the free spectral range (FSR) best suited to the desired application. A schematic of the sensor is presented in **Figure 7**.

The PDMS was cured for 45 min at 338.15 K, fixing the length of the air micro-bubble. Testing was conducted between 324.35–343.65 K and compared to a PT100 thermometer with a resolution of 0.01 K. Results indicated sensitivity was quite high with a value of 2.7035 nm/K and a highly linear response where R^2 = 0.992. In addition to the reported sensitivity, reference was made to the benefit of using a double FPI in the sensor. This consisted of a thin FPI (air micro-bubble) and thick FPI (PDMS filling). The thin FPI allowed for a large FSR and the thick FPI offered high temperature resolution.

Poeggel et al. [34] presented a novel ultra-high resolution temperature sensor (UHRTS). The sensor was comprised of an existing optical fibre pressure temperature sensor (OFPTS) [47] which was further encased in an outer oil-filled capillary. The sensor was noted to have an external diameter of less than 1 mm lending to its capability to be used in volume restricted areas, a schematic of the sensor is provided in **Figure 8**. By combining an FBG with the highly sensitive EFPI, the UHRTS behaved similar to that of Chen et al. [46] in that the FBG and air cavity allowed for a wide spectral range to be utilised with the diaphragm element providing high resolution. Dissimilar to Chen et al. however, the construction of the OFTS was reliant on the thin diaphragm element to transduce volume changes in the oil to a temperature measurement. This being in demonstrated by Eqs. (4) and (5) in Section 3. It was claimed that the high ratio of oil in the outer capillary compared to air in the EFPI cavity was what resulted in high sensitivity. The sensor presented had a reported, theoretical, sensitivity of 52.7×10^3 nm/K. Comparison of the UHRTS to a Bosch BMP085 [48] was carried out over a temperature range of 7 K. While both sensors detected the 7 K variation, measured EFPI sensitivity was much lower than predicted, at 8.77 nm/K. It was suggested that the error was likely caused by inconsistencies in the manufacturing process of the sensor, and the presence of micro air bubbles in the oil which resulted in response damping. Considering oil was introduced into the capillary via a micro-syringe rather than by capillary effect as utilised by several authors such as Chen et al. [46] and Xu et al. [39] this may have

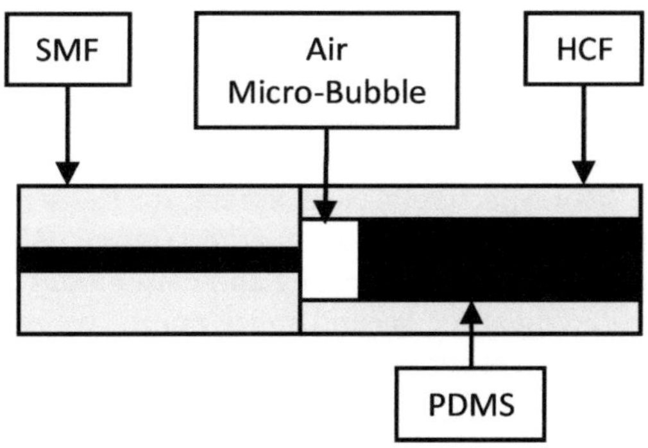

Figure 7.
Schematic of PDMS FPI sensor based on Chen et al. [46].

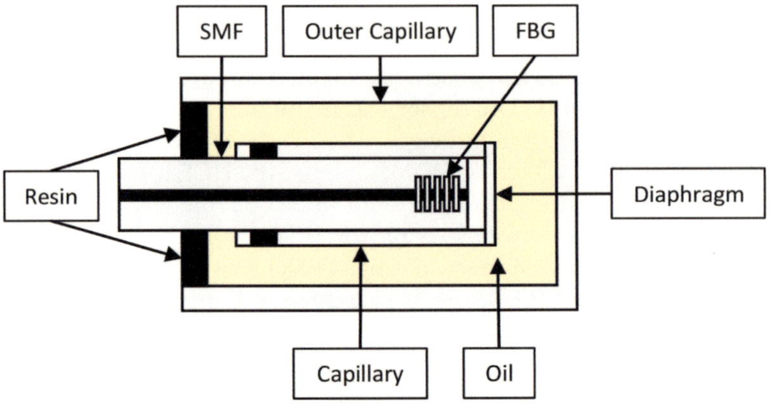

Figure 8.
Schematic of oil-filled EFPI sensor based on Poeggel et al. [34].

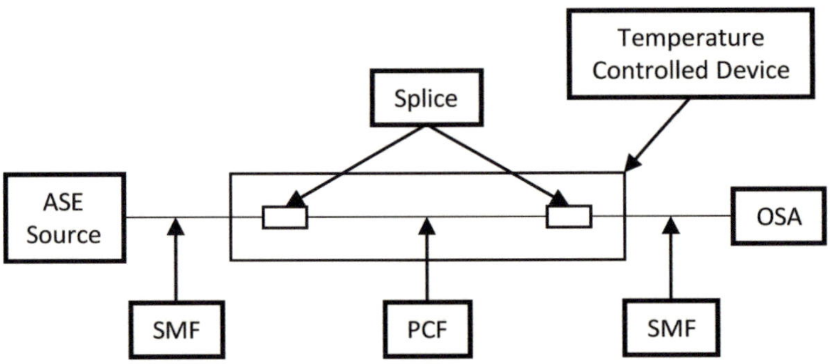

Figure 9.
Schematic of an inline PCF sensor based on Qiu et al. [37].

been a contributing factor to bubble formation. Another potential application of the sensor proposed by Poeggel et al. [34] is use of an ionising radiation sensitive fluid whereby the temperature response varies with exposure to ionising radiation.

4.2 Mach-Zehnder interferometer type sensors

Qiu et al. [37] presented a temperature sensor based on a non-polarimetric PCF, where the sensor was created by splicing a length of commercial single mode (LMA-8) [49] between two lengths of SMF-28. During construction, isopropanol was infused into the non-polarimetric PCF micro-holes by capillary effect, with the PCF subsequently fused between the SMFs. LMA-8 PCF was chosen due to having a diameter equal to that of the SMF-28, this reduced potential complications involved in the splicing process, care was also taken to minimise collapsing of the fibre air holes. Thermo-optic coefficients (TOCs) for the isopropanol and silica are -4.5×10^{-4}/K and 8.6×10^{-6}/K respectively. A schematic for the sensor is provided in **Figure 9**, where an amplified spontaneous emission source (ASE) was used. As light entered from the SMF into the PCF, cladding and core modes propagated at different rates before recombining at the second splice point, this process introduced a phase difference which was observed to be temperature dependant. Akin to the designs of several other fluid-filled inline sensors, this sensor also relied on modifying the core and cladding mode TOCs in order to maximise temperature sensitivity. The sensor was tested in a temperature controlled device over temperatures ranging from 296.85–339.25 K. Blue-shifting of the two tracked waveform dips was observed. A sample of how these waveforms may appear is presented in **Figure 10**. Temperature sensitivities were reported as −133 and − 166 pm/K respectively. While this was an order of magnitude improvement over previously published works [50–52] it was significantly less sensitive when compared to the design of Qian et al. [44], indicating that exploiting the birefringent nature of PCFs may be highly beneficial in the development of inline PCF sensors.

Wang et al. [38] presented a fluid-filled PCF-based modal interferometer (PCFMI). The air holes of the PCF were filled with an oil provided by Cargille Laboratories Inc. (Cedar Grove, NJ, USA) (TOC = -3.37×10^{-4}/K). The system design was similar to that of Qiu et al. [37]. That said however, the operation has similarities to that of Wang et al. [41] with the interference of LP_{01} and LP_{11} modes at the second splicing (recombination) point. Simulation suggested that temperature sensitivity increased proportionally with the ratio of filled to unfilled PCF, and that for constant filling ratio the sensitivity increased with increasing wavelength. It was reported that the latter was due to larger mode field areas of the longer wavelengths. Validation experiments were carried out at three filling ratios (k = 0.256, 0.282, 0.476) over a temperature range of 298.15–355.15 K. Results were in agreement with theoretical prediction, transmission spectra blue-shifted with increasing temperature, and the largest filling ratio (k = 0.476) resulted in the highest temperature sensitivity. Similarly, longer wavelengths resulted in increased sensitivity with a peak value of −340 pm/K at 1480 nm. Another benefit of the proposed sensor type is the linear response to straining, once matrix values were determined, the wavelength shifts may be used to produce temperature and strain measurements simultaneously. The matrix for the sensor presented by Wang et al. is provided by Eq. (7) where S_T and S_ε are the temperature and strain sensitivities respectively. A and B are the two waveform dips which were monitored.

$$\begin{bmatrix} \Delta T \\ \Delta \varepsilon \end{bmatrix} = \begin{bmatrix} S_{T,A} & S_{\varepsilon,A} \\ S_{T,B} & S_{\varepsilon,B} \end{bmatrix}^{-1} \begin{bmatrix} \Delta\lambda_A \\ \Delta\lambda_B \end{bmatrix} \tag{7}$$

Wang et al. [41] presented an ultra-high resolution PCF sensor which had a single liquid filled cladding hole. In the precision filling of the hole, an initial 10 μm end cap was placed on the PCF, after which a hole was precision drilled into the desired PCF hole using a femtosecond laser (FSL). A Cargille Laboratories Inc. immersion oil with a TOC of -3.89×10^{-4} was introduced via capillary effect. The filled region of PCF was subsequently reduced incrementally by ~1 cm until a coherent resonant dip was present. Experimental results between 307.15 and 308.55 K indicated exceptional sensitivity at 54.3 nm/K. Linearity of the results was not provided. Numerical comparison was carried out in Comsol Multiphysics where

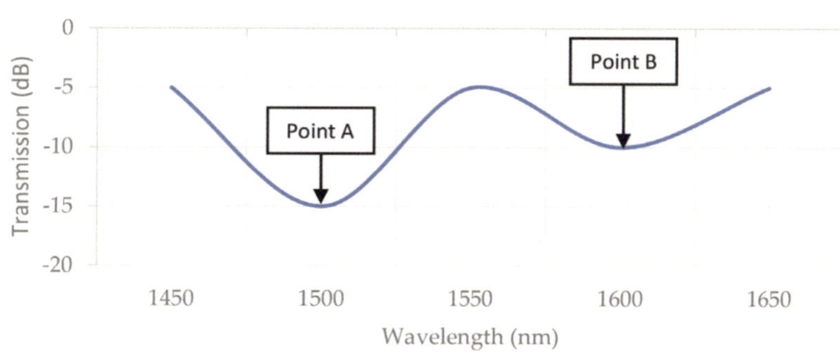

Figure 10.
Representation of how two waveform dips may appear in a transmission spectrum.

the fibre and liquid rod core modes were compared. Results indicated coupling only occurred between the respective LP_{01} modes, where the coupling wavelength shifted by 292 nm between 303.15 and 308.15 K this giving a projected sensitivity of 58.4 nm/K. While predicted sensitivity largely agreed with experimental observations, the location of the dip did not. It was determined that this was predominantly due to uncertainty surrounding the refractive index of the immersion oil, as it had a tolerance of ± 0.002, where an error of 0.001 led to a 150 nm wavelength shift.

Liang et al. [42] reported the first double-filled PCF sensor, with the two fluid rods having varied optical properties, both immersion oils were provided by Cargille Laboratories Inc. The first had a refractive index (RI) of 1.466 with a TOC of -3.91×10^{-4} and the second 1.500 with a TOC of -4.01×10^{-4} respectively. Owing to energy differences the PCF LP_{11} core mode was neglected. Finite element analysis indicated an interaction existed between LP_{01}(core)–LP_{01}(rod 1), and LP_{01} (core)–LP_{11}(rod 2). This indicated two waveform dips would be present in the transmission spectrum. Furthermore rod 1 displayed red-shift with increasing temperature with the converse being true of rod 2. As the two liquid rods were relatively far apart geometrically, there was no reported interaction. Temperature response was recorded between 325.15 and 327.15 K in increments of 0.2 K. Dip sensitivities were recorded as being 42.818 and − 11.343 nm/K with linearities of R^2 = 0.99951 and 0.99935, thus indicating the sensor had extremely high sensitivity. A highly sensitive strain response was also reported. Force on the fibre was increased from 0.218 to 0.855 N in increments of 0.049 N, with strain sensitivities being −38.041 and 8.702 pm/$\mu\varepsilon$ Linearities were R^2 = 0.99869 and 0.99495.

Xu et al. [39] utilised a thin core fibre (TCF) rather than PCF in the development of their sensor. By immersing the TCF in a Cargille Laboratories Inc. immersion oil and sealing the PCF within a capillary, the influences of external refractive indices was eliminated. The TCF was approximately 20 mm long with the protective capillary 40 mm long. A schematic of the sensor is given in **Figure 11**. Similar to PCF style temperature sensors, there was a resonant dip in the transmission spectrum, which was located at 1561.7 nm for the unfilled sensor. Temperature was modulated on an unspecified thermoelectric cooler, which could be controlled to 0.1 K resolution in the temperature range 288.15–318.15 K. Results were in good agreement with other published works, where a temperature sensitivity of 9.0 pm/K was recorded with a sensor linearity of R^2 = 0.9957.

The introduction of immersion oil (TOC = −3.95 × 10^{-4}/K) and capillary resulted in higher temperatures moving the resonance dip to a shorter wavelength, i.e. blue-shift, this being the converse of the bare TCF Mach-Zehnder Interferometer (TCFMI). Experimental results indicated that introduction of the immersion oil

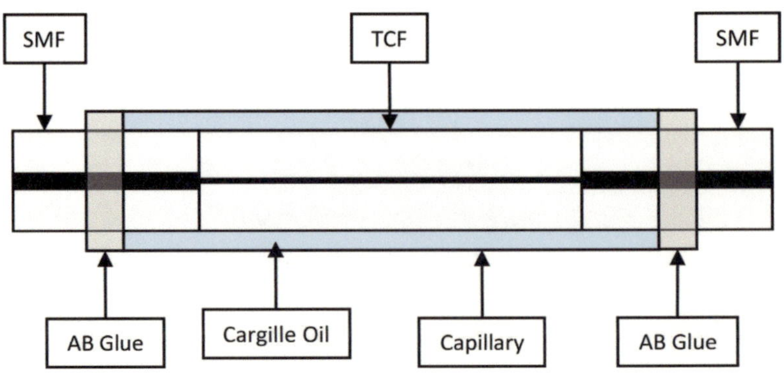

Figure 11.
Schematic of sensor based on Xu et al. [39].

as a sealing fluid increased the temperature sensitivity of the interferometer >250× compared to the baseline at −2.3 nm/K. Heating and cooling the sensor showed repeatable results with the fitted polynomial having an R^2 = 0.999. It should be noted however at higher temperatures, the sensitivity decreased and the fitting error increased. Peak resolution was reported at 283.15 K as 0.008 K with an OSA resolution of 0.2 pm.

Xu et al. [39] went further to quantify the influence of strain on the sensor by measuring the temperature responses of a mechanically strained bare TCFMI, and an air-sealed TCFMI. Results indicated that the resonance-dip blue shifted with strain on the bare TCFMI, and the air sealed TCFMI appeared relatively temperature insensitive. The work concluded on the point of the sensitivity being primarily driven by the sealing fluid's thermo-optic coefficient. Again, however, the sensor required transmission of the light in order to be used, thus eliminating its potential to be used as a point sensor.

4.3 Sagnac interferometer type sensors

Qian et al. [44] presented an alcohol filled temperature sensor based on a highly birefringent photonic crystal fibre (HiBi-PCF) within a fibre loop mirror (FLM) as illustrated in **Figure 12**, the light source used was a super-luminescent light emitting diode (SLED). Owing to the birefringence of the HiBi-PCF, the counter-propagating waves caused by the 3 dB coupler have an optical path difference at recombination. Two resonant dips manifested in the transmission spectrum at 293.15 K, these being present at 1455.8 and 1549.8 nm. The sensor was tested in two conditions within an unspecified controlled temperature chamber. The first of these increasing from 293.15 to 307.15 K and the latter reducing the temperature from 293.15 to 281.15 K with the two resonant dips' responses being monitored. Measurement linearities were R^2 = 0.9995 and R^2 = 0.9997 respectively. Measured sensitivities were 6.2 and 6.6 nm/K compared to the theoretical values of 6.1 and 6.5 nm/K.

Cui et al. [45] proposed an SI type sensor similar in construction to that of Qian et al. [44]. The study conducted, however, went further in an effort to quantify the influence of selective hole filing in the PCF versus non-selective filling. Further to this, the length of PCF and hole fill ratio were explored. Simulations of no infiltration, all holes filled, small holes filled, and big holes filled were carried out; with the birefringence sensitivity to infiltrating liquid being monitored. While all three liquid filled cases indicated a reduced PCF birefringence, the 'big holes filled'

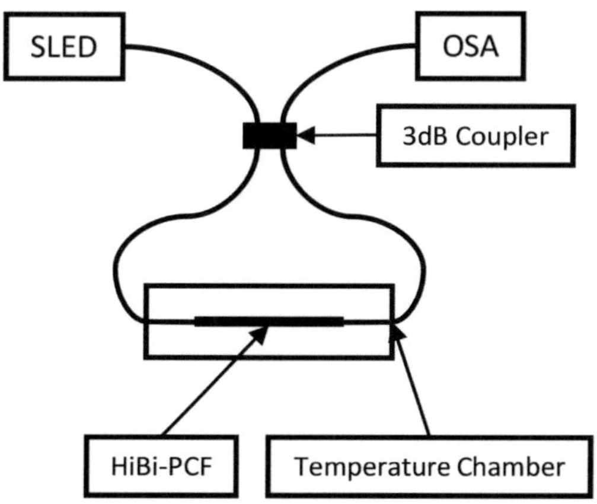

Figure 12.
Schematic of HiBi-PCF sensor based on Qian et al. [44].

condition resulted in the highest temperature sensitivity, with a birefringence change of 27% as liquid refractive index was varied from 1.33 to 1.36. Dissimilar to the selective collapsing and cleaving method employed by Peng et al. [40] Cui et al. [45] offered a simplistic method of sealing the outer holes by introducing a micro-droplet of glue to the fibre face while monitoring with a microscope. It was claimed that the process could be conducted in under a minute with repeatable results after minimal training. During experimentation, water was used in place of ethanol due to the high coefficient of thermal expansion, and reduced tendency to evaporate. The sensor indicated a good sensitivity of −2.58 nm/K with a linearity of R^2 = 0.9991. The OSA used to conduct the experiment had a resolution of 0.02 nm thus giving the sensor a resolution in the region of 7.75 mK. Accounting for the length of liquid filled PCF, results similar to that of Wang et al. [38] indicated increasing the ratio of filled to unfilled PCF increased temperature sensitivity.

4.4 Similar sensors

While one could consider liquid crystal based sensors in a similar category to those above, the authors wish to express they fall outside the remit of this chapter. However should the reader wish to explore these technologies, Windhorn and Cain [25] provides a good starting point. Several recent works exploring the use of liquid crystal thermometry using PCFs have also been published [53–55]. Wolinski et al. presented a multi-faceted liquid crystal PCF which was capable of temperature, electric field, and hydrostatic pressure detection [56].

5. Conclusions

The remit of this review was to address current state-of-the-art LiF-OFTSs presented in literature. The optical fibre sensors examined were based on the principles of extrinsic Fabry-Perot interferometers, Mach-Zehnder interferometers and Sagnac interferometers. While sensitivities varied to an extent, they each presented performance equivalent to, or better than, reported electrical sensors. As noted in other reviews of optical fibre sensing such as that of Lee [57] and Poeggel et al. [58] optical fibre sensors had numerous additional advantages over their electronic counterparts such as immunity to EMI, small form factor, along with distributed and multiplexed sensing capability. Perhaps encouragingly, the commercial marketplace for optical fibre sensors appears to have matured significantly since the publication of Lee's review [57]. A driving factor in this will likely have been the coupled high resolution and fast response times, which are generally sub-Kelvin and sub-second respectively.

While FBGs remain the predominant method of temperature sensing, EFPI sensors have gained momentum in the commercial arena with companies such as FISO & Opsens offering varied ranges, sensitivities and sampling rates, indicating that they are slowly gaining favour in engineering and biomedical applications. That said however, MZI based temperature sensors were the more reported sensor type in literature. From this review it was obvious that while several ultra-high resolution liquid filled temperature sensors have been developed there is still scope for significant work to be carried out to improve their performance and stimulate widespread commercial deployment.

Considering that a significant portion of optical fibre sensor research has been focused towards biomedical applications in recent years, it is understandable that a market for inline sensors may not exist, that said however it may bode well for the

Company	Proximion	OPSENS	RJC Enterprises
Sensor name	WISTHEAT [60]	OTG-MPK5 [61]	N/A [62]
Sensor type	FBG	GaAs Crystal	EFPI
Min. temperature	228.15 K	293.15 K	288.15 K
Max. temperature	523.15 K	318.15 K	328.15 K
Resolution	70 mK(0.5 pm OSA)	10 mK	100 mK

Table 2.
Commercially available optical fibre temperature sensors.

designs of Chen et al. [46] and Poeggel et al. [34]. Another factor which has often been overlooked is the filling liquid properties, the majority of MZI sensors used an immersion oil provided by Cargille Laboratories Inc. and one reported using isopropanol. While the isopropanol may act as an irritant, the immersion oil used may be toxic if swallowed or inhaled, such as that of the Series AA [59]. It is also known to be damaging to waterways, thus indicating strict environmental controls require consideration. That said however, monitoring of industrial equipment using this method is more than plausible with the potential to use multiple PCFs on a single fibre to provide distributed sensing, **Table 2** lists example commercially available OFTSs. While the sensors are not of a liquid filled construction, they indicate the minimum required performance of any potential liquid filled sensor in order to potentially be commercially competitive.

Acknowledgements

This work was supported by the Science Foundation Ireland grant number: 15/CDA/3598

Conflict of interest

The authors wish to declare no conflict of interest.

Nomenclature

Symbol	Name	Unit
D	Flexural rigidity	$Pa.m^3$
E	Young's modulus	N/m^2
P	Pressure	N/m^2
Q	Heat	J
R	Outer radius	m
S_T	Temperature sensitivity	/K
S_ε	Strain sensitivity	-
T	Temperature	K
T_0	Initial temperature	K
c	Specific heat capacity	J/kg.K
h	Thickness	m
n_{eff}	Effective refractive index	-
r	Reference radius	m

t	Time	s
z	Deflection	m
Λ	Bragg grating pitch	m
α_n	Thermal coefficient of refractive index	/K
α_d	Thermal coefficient of thickness	/K
ε	Strain	-
λ	Wavelength	m
λ_B	Bragg wavelength	m
ν	Poisson's ratio	-

Author details

Fintan McGuinness[1], Gabriel Leen[1], Elfed Lewis[1], Gerard Dooly[2], Daniel Toal[2] and Dinesh Babu Duraibabu[2*]

1 Optical Fibre Sensors Research Centre, University of Limerick, Limerick, Ireland

2 Centre for Robotics and Intelligent Systems, University of Limerick, Limerick, Ireland

*Address all correspondence to: dineshbabu.duraibabu@ul.ie

References

[1] Valleriani M. Galileo Engineer. In: Cohen RS, Renn J, Gavroglu K, editors. Boston Studies in the Philosophy of Science. Vol. 269. Dordrecht: Springer Science + Business Media; 2010. pp. 155-165

[2] McGee TD. Principles and Methods of Temperature Measurement. 1st ed. New York: Wiley-Interscience; 1988

[3] Sverdrup H, Johnson MW, Fleming RH. The Oceans, Their Physics, Chemistry, and General Biology. New York: Prentice-Hall Inc.; 1942 [Internet]. Available from: http://ark.cdlib.org/ark:/13030/kt167nb66r/

[4] Sarma U, Boruah PK. Design and development of a high precision thermocouple based smart industrial thermometer with on line linearization and data logging feature. Measurement: Journal of the International Measurement Confederation. 2010; **43**(10):1589-1594. DOI: 10.1016/j.measurement.2010.09.003 [Internet]

[5] Yang F, Li G, Yang J, Wang Z, Han D, Zheng F, et al. Measurement of local temperature increments induced by cultured HepG2 cells with micro-thermocouples in a thermally stabilized system. Scientific Reports. 2017;**7**(1): 1-12. DOI: 10.1038/s41598-017-01891-1 [Internet]

[6] Resistor Guide, NTC thermistor Resistor Guide [Internet]. Available from: http://www.resistorguide.com/ntc-thermistor/ [Accessed: 25-07-2018]

[7] Clow GD, Saltus RW, Waddington ED. A new high-precision borehole-temperature logging system used at GISP2, Greenland, and Taylor Dome, Antarctica. Journal of Glaciology. 1996; **42**(142):576-584. DOI: 10.3189/S0022143000003555 [Internet]. Available from: https://www.cambridge.org/core/product/identifier/S0022143000003555/type/journal%7B%5C_%7Darticle

[8] Hanna Instruments, Ultra-Fast Penetration K-Type Thermocouple Probe with Handle [Internet]. 2018. Available from: https://hannainst.com/hi766c1-ultra-fast-penetration-%20k-type-thermocouple-probe-with-handle.html [Accessed: 27-06-2018]

[9] Hammond DL, Benjaminson A. The linear quartz thermometer—A new tool for measuring absolute and difference temperatures. Hewlett-Packard Journal. 1965;**16**(7):1-7 [Internet]. Available from: ftp://dns.soest.hawaii.edu/bhowe/outgoing/20120920%7B%5C_%7DITU%7B%5C_%7DParis/HPJ-1965-03.pdf

[10] Lerch BA, Nathal MV. Thermocouple Calibration and Accuracy in a Materials Testing Laboratory. TM-2002-211507. Ohio: Glenn Research Centre, NASA; 2002

[11] IEEE Standards Association. IEEE Std 2700™-2014, IEEE Standard for Sensor Performance Parameter Definition. New York: IEEE; 2014. DOI: 10.1109/70 IEEESTD.2014.6880296

[12] Rhoades R, Pflanzer R. In: Bachman C, editor. Human Physiology. 4th ed. Pacific Grove: Thomson Learning Inc.; 2003

[13] Omega. TCC type correltaion. Tech. Rep. type C [Internet]. pp. 4-5. Available from: https://www.omega.com/techref/pdf/z202.pdf

[14] Buckton G, Russell SJ, Beezer AE. Pharmaceutical calorimetry: A selective review. Thermochimica Acta. 1991;**193** (C):195-214. DOI: 10.1016/0040-6031(91)80184-K

[15] Rossi S. Brain temperature, body core temperature, and intracranial

pressure in acute cerebral damage. Journal of Neurology, Neurosurgery & Psychiatry. 2001;**71**(4):448-454. DOI: 10.1136/jnnp.71.4.448 [Internet]

[16] NASA. Heat Transfer [Internet]. 2018. Available from: https://wright.nasa.gov/airplane/heat.html [Accessed: 12-07-2018]

[17] Burgett RR, Massoll RE, Van Uum DR. Relationship between spark plugs and engine-eadiated electromagnetic interference. IEEE Transactions on Electromagnetic Compatibility. 1974; **EMC-16**(3):160-172. DOI: 10.1109/TEMC.1974.303355

[18] GE Oil & Gas. Platinum Resistance Thermometer (PRT) [Internet]. 2012. Available from: https://geoilandgas.com/sites/geog/files/prt%7B%5C_%7Dplatinum%7B%5C_%7Dresistance%7B%5C_%7Dthermometer.pdf [Accessed: 27-06-2018]

[19] Hanna Instruments. K-Thermometers [Internet]. 2018. Available from: https://hannainst.com/downloads/dl/file/id/1637/manhi%7B%5C_%7D93530%7B%5C_%7D531%7B%5C_%7D532.pdf [Accessed: 27-06-2018]

[20] Shenzhen Senstech Electronic Technology Company Ltd. Waterproof Temperature Sensor DS18B20 Programmable Resolution 1-Wire Digital Thermometer [Internet]. 2018. Available from: https://www.alibaba.com/product-detail/Waterproof-Temperature-Sensor-DS18B20-Programmable-%20Resolution%7B%5C_%7D1898805574.html?spm=a2700.7724857.main07.31.40cb3892YwvGzX [Accessed: 27-06-2018]

[21] Castrellon J, Paez G, Strojnik M. Remote temperature sensor employing erbium-doped silica fiber. Infrared Physics and Technology. 2002;**43**(3–5):219-222. DOI: 10.1016/S1350-4495(02)00142-1

[22] Zhang H, Jiang Q, Wang BY, Wang JJ. Monitoring diesel engine parameters based on FBG probe. Optoelectronics Letters. 2016;**12**(5):0384-0388. DOI: 10.1007/s11801-016-6162-7

[23] Pinet É, Hamel C, Glišić B, Inaudi D, Miron N. Health monitoring with optical fiber sensors: from human body to civil structures. In: Kundu T, editor. SPIE Proceedings. San Diego: SPIE; 19 May 2007;**6532**:653219-653230. DOI: 10.1117/12.715186. [Internet]. Available from: http://proceedings.spiedigitallibrary.org/proceeding.aspx?articleid=1337730%20; http://proceedings.spiedigitallibrary.org/proceeding.aspx?doi=10.1117/12.715186

[24] Corning Incorporated. Frequently Asked Questions on Fiber Reliability [Internet]. 2016. Available from: https://www.corning.com/media/worldwide/coc/documents/Fiber/RC-%20White%20Papers/WP5082%203-31-2016.pdf

[25] Windhorn TH, Cain CA. Optically active binary liquid crystal thermometry. IEEE Transactions on Bio-Medical Engineering. 1979;**26**(3):148-152 [Internet]. Available from: http://www.ncbi.nlm.nih.gov/pubmed/521025

[26] Poeggel S, Tosi D, Duraibabu D, Kelly J, Munroe M, Leen G, et al. Novel diaphragm microfabrication techniques for high-sensitivity biomedical fiber optic Fabry-Perot interferometric sensors. In: Proceedings of SPIE—The International Society for Optical Engineering. Vol. 9098. 2014. p. 909813. DOI: 10.1117/12.2050500 [Internet]. Available from: http://proceedings.spiedigitallibrary.org/proceeding.aspx?doi=10.1117/12.2050500

[27] Nakazaki Y, Yamashita S. Fast and wide tuning range wavelength-swept fiber laser based on dispersion tuning and its application to dynamic FBG

sensing. Optics Express. 2009;**17**(10): 8310. DOI: 10.1364/OE.17.008310 [Internet]. Available from: https://www.osapublishing.org/abstract.cfm?URI=oe-17-10-8310

[28] Poeggel S, Duraibabu D, Tosi D, Leen G, Lewis E, McGrath D, et al. Differential *in vivo* urodynamic measurement in a single thin catheter based on two optical fiber pressure sensors. Journal of Biomedical Optics. 2015;**20**(3):037005. DOI: 10.1117/1.JBO.20.3.037005 [Internet]. Available from: http://biomedicaloptics.spiedigitallibrary.org/article.aspx?doi=10.1117/1.JBO.20.3.037005

[29] Sellnau MC, Matekunas FA, Battiston PA, Chang C-F, Lancaster DR. Cylinder-Pressure-Based Engine Control Using Pressure-Ratio-Management and Low-Cost Non-Intrusive Cylinder Pressure Sensors no. 724 [Internet] . 2000. DOI: 10.4271/2000-01-0932. Available from: http://papers.sae.org/2000-01-0932/

[30] Kasper JC, Wiggenhorn M, Resch M, Friess W. Implementation and evaluation of an optical fiber system as novel process monitoring tool during lyophilization. European Journal of Pharmaceutics and Biopharmaceutics. 2013;**83**(3):449-459. DOI: 10.1016/j.ejpb.2012.10.009 [Internet]

[31] Hong CY, Zhang YF, Zhang MX, Leung LMG, Liu LQ. Application of FBG sensors for geotechnical health monitoring, a review of sensor design, implementation methods and packaging techniques. Sensors and Actuators, A: Physical. 2016;**244**:184-197. DOI: 10.1016/j.sna.2016.04.033 [Internet]

[32] Gassino R, Pogliano J, Perrone G, Vallan A. Issues and characterization of fiber Bragg grating based temperature sensors in the presence of thermal gradients. Measurement: Journal of the International Measurement Confederation. 2018;**124**(January): 15-19. DOI: 10.1016/j.measurement.2018.03.049 [Internet]

[33] Benham P, Crawford R, Armstrong C. Mechanics of Engineering Materials. 2nd ed. Harlow: Prentice-Hall Inc.; 1996

[34] Poeggel S, Duraibabu D, Dooly G, Lewis E, Leen G. Novel ultrahigh resolution optical fibre temperature sensor. In: Lewis E, editor. Proceedings of SPIE—The International Society for Optical Engineering. Vol. 9916. 2016. p. 991605. DOI: 10.1117/12.2236155 [Internet]. Available from: http://proceedings.spiedigitallibrary.org/proceeding.aspx?doi=10.1117/12.2236155

[35] Gao X, Yang M, Peng J, Lv D. Miniature fiber-optic temperature sensor based on optical coating interference. Optik—International Journal for Light and Electron Optics. 2017;**130**:1014-1020. DOI: 10.1016/j.ijleo.2016.11.114 [Internet]. Available from: http://linkinghub.elsevier.com/retrieve/pii/S0030402616314516

[36] Hecht E. Optics. 5th ed. Harlow: Pearson Education Limited; 2014

[37] Qiu S-j, Chen Y, Xu F, Lu Y-q. Temperature sensor based on an isopropanol-sealed photonic crystal fiber in-line interferometer with enhanced refractive index sensitivity. Optics Letters. 2012;**37**(5):863-865. DOI: 10.1364/OL.37.000863 [Internet]. Available from: http://www.ncbi.nlm.nih.gov/pubmed/22378419

[38] Wang S, Liu Y-g, Wang Z, Han T, Xu W, Wang Y, et al. Intermodal interferometer based on a fluid-filled two-mode photonic crystal fiber for sensing applications. Applied Optics. 2013;**52**(14):3166-3171. DOI: 10.1364/AO.52.003166 [Internet]. Available from: http://ao.osa.org/abstract.cfm?URI=ao-52-14-3166

[39] Xu B, Li J, Li Y, Xie J, Dong X. Liquid seal for temperature sensing with

fiber-optic refractometers. Sensors (Switzerland). 2014;**14**(8):14873-14884. DOI: 10.3390/s140814873

[40] Peng Y, Hou J, Zhang Y, Huang Z, Xiao R, Lu Q. Temperature sensing using the bandgap-like effect in a selectively liquid-filled photonic crystal fiber. Optics Letters. 2013;**38**(3): 263-265. DOI: 10.1364/OL.38.000263 [Internet]. Available from: http://www.ncbi.nlm.nih.gov/pubmed/23381405

[41] Wang Y, Yang M, Wang DN, Liao CR. Selectively infiltrated photonic crystal fiber with ultrahigh temperature sensitivity. IEEE Photonics Technology Letters. 2011;**23**(20):1520-1522

[42] Liang H, Zhang W, Geng P, Liu Y, Wang Z, Guo J, et al. Simultaneous measurement of temperature and force with high sensitivities based on filling different index liquids into photonic crystal fiber. Optics Letters. 2013;**38**(7): 1071-1073. DOI: 10.1364/OL.38.001071 [Internet]. Available from: http://www.ncbi.nlm.nih.gov/pubmed/23546247

[43] Napolitano F. Fiber-Optic Gyroscopes Key Technological Advantages. Saint-Germain en Laye, Tech. Rep. [Internet]. 2010. Available from: https://www.ixblue.com/m/publication/fog-key-advantages.pdf

[44] Qian W, Zhao C-L, He S, Dong X, Zhang S, Zhang Z, et al. High-sensitivity temperature sensor based on an alcohol-filled photonic crystal fiber loop mirror. Optics Letters. 2011;**36**(9):1548. DOI: 10.1364/OL.36.001548. arXiv: 0604155 [physics]. [Internet]. Available from: https://www.osapublishing.org/abstract.cfm?URI=ol-36-9-1548

[45] Cui Y, Shum PP, Hu DJJ, Wang G, Humbert G, Dinh XQ. Temperature sensor by using selectively filled photonic crystal fiber Sagnac interferometer. IEEE Photonics Journal. 2012;**4**(5):1801-1808. DOI: 10.1109/JPHOT.2012.2217945

[46] Qing Chen M, Zhao Y, Xia F, Peng Y, jie Tong R. High sensitivity temperature sensor based on fiber air-microbubble Fabry-Perot interferometer with PDMS-filled hollow-core fiber. Sensors and Actuators, A: Physical. 2018;**275**:60-66. DOI: 10.1016/j.sna.2018.03.044 [Internet]

[47] Poeggel S, Duraibabu D, Kalli K, Leen G, Dooly G, Lewis E. Recent improvement of medical optical fibre pressure and temperature sensors. Biosensors. 2015;**5**(3):432-449. DOI: 10.3390/bios5030432 [Internet]. Available from: http://www.mdpi.com/2079-6374/5/3/432/

[48] Sensortec B. Data sheet. Tech. Rep. [Internet]. 2009. Available from: https://www.sparkfun.com/datasheets/Components/General/BST-BMP085-DS000-05.pdf

[49] NKT Photonics. LMA-8. Tech. Rep. [Internet]. Available from: https://www.nktphotonics.com/wp-content/uploads/sites/3/2015/01/LMA-8.pdf

[50] Jha R, Villatoro J, Badenes G, Pruneri V. Refractometry based on a photonic crystal fiber interferometer. Optics Letters. 2009;**34**(5):617. DOI: 10.1364/OL.34.000617 [Internet]. Available from: https://www.osapublishing.org/abstract.cfm?URI=ol-34-5-617

[51] Qiu S-j, Chen Y, Kou J-l, Xu F, Lu Y-q. Miniature tapered photonic crystal fiber interferometer with enhanced sensitivity by acid microdroplets etching. Applied Optics. 2011;**50**(22):4328. DOI: 10.1364/AO.50.004328 [Internet]. Available from: https://www.osapublishing.org/abstract.cfm?URI=ao-50-22-4328

[52] Gong H, Chan CC, Zhang YF, Wong WC, Dong X. Miniature refractometer based on modal interference in a hollow-core photonic crystal fiber with

collapsed splicing. Journal of Biomedical Optics. 2011;**16**(1):017004. DOI: 10.1117/1.3527259 [Internet]. Available from: http://biomedicaloptics.spiedigitallibrary.org/article.aspx?doi=10.1117/1.3527259

[53] Wolinski TR, Szaniawska K, Ertman S, Lesiak P, Domanski AW, Dabrowski R, et al. Influence of temperature and electrical fields on propagation properties of photonic liquid-crystal fibres. Measurement Science and Technology. 2006;**17**(5):985-991. DOI: 10.1088/0957-0233/17/5/S08 [Internet]. Available from: http://stacks.iop.org/0957-0233/17/i=5/a=S08?key=crossref.f9489a53384ab98a6cd800efb7324542

[54] Wei L, Eskildsen L, Weirich J, Scolari L, Alkeskjold TT, Bjarklev A. Continuously tunable all-in-fiber devices based on thermal and electrical control of negative dielectric anisotropy liquid crystal photonic bandgap fibers. Applied Optics. 2009;**48**(3):497. DOI: 10.1364/AO.48.000497 [Internet]. Available from: https://www.osapublishing.org/abstract.cfm?URI=ao-48-3-497

[55] Miao Y, Liu B, Zhang K, Liu Y, Zhang H. Temperature tunability of photonic crystal fiber filled with Fe_3O_4 nanoparticle fluid. Applied Physics Letters. 2011;**98**(2):21103. DOI: 10.1063/1.3540647

[56] Wolinski T, Czapla A. Photonic liquid crystal fibers for sensing applications. IEEE Transactions on Instrumentation and Measurement. 2008;**57**(8):1796-1802. DOI: 10.1109/TIM.2008.922077 [Internet]. Available from: http://ieeexplore.ieee.org/xpls/abs%7B%5C_%7Dall.jsp?arnumber=4510764

[57] Lee B. Review of the present status of optical fiber sensors. Optical Fiber Technology. 2003;**9**(2):57-79. DOI: 10.1016/S1068-5200(02)00527-8

[58] Poeggel S, Tosi D, Duraibabu D, Leen G, McGrath D, Lewis E. Optical fibre pressure sensors in medical applications. Sensors. 2015;**15**(7):17115-17148. DOI: 10.3390/s150717115 [Internet]. Available from: http://www.mdpi.com/1424-8220/15/7/17115/

[59] Cargille Laboratories Inc. Cargille Refractive Index Liquid Series AA nD = 1.400–1.459 Safety Data Sheet [Internet]. Available from: http://www.cargille.com/SDS%7B%5C_%7DRI%7B%5C_%7DSeries%7B%5C_%7DAA.pdf

[60] Proximion. High Temperature FBG Sensor for Harsh Environment [Internet]. Available from: https://static1.squarespace.com/static/54690ca0e4b0dc0a73f1772b/t/57f752ed579fb360231dc1f4/1475826414976/Proximion%7B%5C_%7Dprodblad%7B%5C_%7DWISTHEAT%7B%5C_%7D161005.pdf [Accessed: 12-07-2018]

[61] OpSens. OTG-MPK5 Fiber Optic Temperature Sensor, Probe and Transducer [Internet]. 2018. Available from: https://opsens-solutions.com/products/fiber-optic-temperature-sensors/otg-mpk5/ [Accessed: 12-07-2018]

[62] RJC Enterprises. How Our Sensors Work [Internet] 2018. Available from: http://www.rjcenterprises.net/background.html [Accessed: 12-07-2018]

Chapter 7

Optical Fibre Long-Period Grating Sensors Operating at and around the Phase Matching Turning Point

Rebecca Yen-Ni Wong, Dora Hu Juan Juan, Morten Ibsen and Perry Ping Shum

Abstract

Optical fibres have been exploited as sensors for many years and they provide a versatile platform with a small form factor. Long-period gratings (LPGs) operating at and around the phase matching turning point (PMTP) possess some of the highest sensitivities to external perturbations in the family of LPG-based sensor devices. This type of optical fibre grating has been demonstrated as a sensor for use in a wide range of applications. In this review chapter, an overview of PMTP LPGs is presented and the key developments, findings and applications are highlighted. The fabrication considerations and sensor limitations are also discussed.

Keywords: optical fibre, fibre optics, fibre sensors, long-period gratings, phase matching turning point, turn around point

1. Introduction

Optical fibre sensors do not only have use in telecommunications but are also extremely useful in a number of sensing applications. Many fields such as medical, oil and gas, civil, automotive as well as aerospace industries (structural health monitoring) have benefitted from optical fibre grating sensors [1–4].

In-fibre gratings are known as intrinsic sensing devices and therefore the propagation of light is guided and controlled within the fibre. Fibre gratings have a perturbation with a certain periodicity which will cause the fibre properties to change. They are also relatively easy to configure, are wavelength encoded enabling stable signals, and offer a high signal-to-noise ratio. One type of in-fibre grating is the long-period grating (LPG), which Vengsarkar et al. [5, 6] were the first to introduce. LPGs typically have periods ranging from around 100 μm to around 1 mm [7]. The principle of operation consists of the forward propagating core mode coupling with one or more of the forward propagating cladding modes [8]. The coupling involves the cladding modes, which means that the evanescent field will extend into the fibre surroundings. This will cause the LPG to be affected by its local environment. Another type of in-fibre grating is the fibre Bragg grating (FBG). The FBG promotes coupling of the propagating core mode with the counter-propagating core mode. FBGs typically have sub-micron periods and will produce a peak (in reflection) at a wavelength that is able to satisfy the Bragg condition. FBGs have also been used for numerous sensing applications [9, 10], but they will not be covered in this chapter.

By appropriately selecting the period of an LPG, it is possible to ensure the core mode will couple to a cladding mode operating at the turn around point (TAP) [11], also known as the phase matching turning point (PMTP), or dispersion turning point (DTP). A feature known as the dual resonance band can also be produced in this region. This type of LPG configuration has become increasingly popular due to its ultra-high sensitivity, a property usually desirable for a sensor. Approaches employed to improve the sensing capability of LPGs have included methods such as tapering [12] and etching [13]; however this can weaken the structure of the fibre and requires more delicate handling or complicated packaging. These sensors have been successfully used for measuring parameters such as temperature [14–16], strain [14–16] and refractive index (RI) [17–20]. The properties of LPGs at PMTP can be tailored further by adding a functional nanoscale coating for chemical and gas sensing [21]. This enables users to adapt the sensor to their own needs and applications. Chemical and bio-chemical based sensors, or those that can be applied to healthcare, are attracting increasing attention as they can have a more direct impact on the wellbeing of people. However, many are still yet to be applied in real situations outside of the laboratory [2].

This chapter aims to provide a more comprehensive coverage of LPGs which operate at and around the phase matching turning point, with respect to what can be found in existing literature [22]. The typical characteristics and fabrication considerations will be discussed. This will be followed by the different applications where PMTP LPGs have been demonstrated.

2. Long-period gratings at phase matching turning point

LPGs consist of some periodic modulation in the optical fibre which causes the core mode to couple with a number of modes in the cladding, at discrete wavelengths (**Figure 1**). The modes will propagate through the fibre with the propagation constants, β_{co} *and* β_{cl}^{M} (core and *M*-th cladding mode, respectively) and the wavelengths are dependent upon the satisfaction of the phase matching condition, which is described as [16]:

$$\lambda = [n_{co}(\lambda) - n_{cl}^{M}(\lambda)]\Lambda \quad (1)$$

Where λ is the resonant wavelength, n_{co} is the effective refractive index of the propagating core mode, n_{cl}^{M} is the effective refractive index of the *M*-th cladding mode, and the Λ is the period of the LPG.

Light that couples to the fibre cladding modes is lost rapidly through scattering and absorption at the cladding and the surrounding medium interface. This is presented as a transmission spectrum, as shown in **Figure 2**, that contains one or more resonance bands with wavelengths, λ_n.

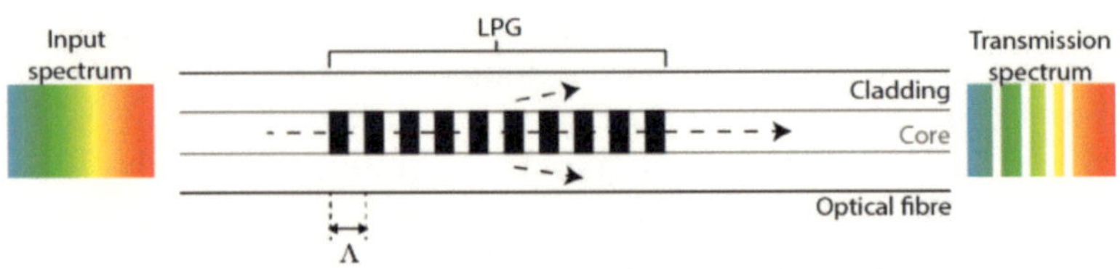

Figure 1.
Schematic diagram of an LPG.

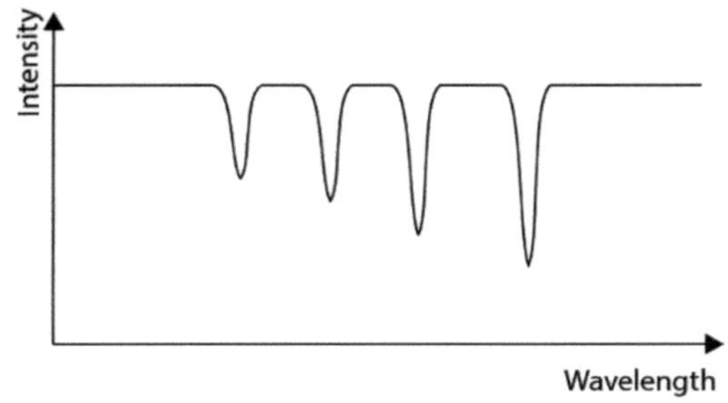

Figure 2.
Illustration of an LPG transmission spectrum with resonance bands at discrete wavelengths.

Eq. (1) shows the resonant wavelength is dependent on the effective refractive indices of the core and cladding mode of the fibre. This central resonant wavelength and the sensitivity of an LPG is affected by the order of the coupled cladding mode; choosing the grating period, the type [23] and composition of the fibre [19], and any external perturbations can alter the coupled mode.

Optical fibre LPG structures (a three layer cylindrical waveguide consisting of the core, cladding and ambient surrounding) can be modelled using coupled mode theory [24–26]. This can be used to describe the power transmitted between the modes of the waveguides. Guided modes propagating in a fibre can be treated as linearly polarised (LP), employing the weakly guided approximation [27]; this approximates the difference between the normalised core and cladding refractive index, Δ, to be very small [24, 25, 27]:

$$\Delta = \frac{n_{co} - n_{cl}}{n_{co}} \ll 1 \tag{2}$$

The coupled mode theory equations that are used to describe the LPG can be simplified to [24]:

$$\frac{dA_{co}}{dz} = i\,k_{co-co} + i\sum_{v} \frac{m}{2}\, k_{cl-co}^{M}\, A_{cl}^{M}\, e^{\left(-i\,2\,\delta_{cl-co}^{M}\, z\right)}, \tag{3}$$

$$\sum_{v} \frac{dA_{cl}^{M}}{dz} = +i\,\frac{m}{2}\, k_{cl-co}^{M}\, A_{co}\, e^{\left(+i\,2\,\delta_{cl-co}^{M}\, z\right)} \tag{4}$$

Where A_{co} is the amplitude of the core mode along the z-axis, A_{cl}^{M} is the amplitude of the cladding mode along the z-axis, the z-axis is along the axis of the optical fibre, k is the coupling constant, m is the induced-index fringe modulation.

The small-detuning factor for the co-propagating modes is defined as:

$$\delta_{cl-co}^{M} \equiv \frac{1}{2}\left(\beta_{co} - \beta_{cl}^{M} - \frac{2\pi}{\Lambda}\right) \tag{5}$$

Phase matching curves of resonance wavelength against grating period of an LPG can be generated by calculating the dispersion of the modes of the core and the cladding. These sets of curves are able to predict coupling from the core to the cladding mode, and that for each cladding mode there will be a turning point [28]. Around the turning point, a single mode can be coupled at two different wavelengths simultaneously [11, 28]. **Figure 3** shows an example of phase matching curves for higher order cladding modes in the 600–1150 nm wavelength range. It

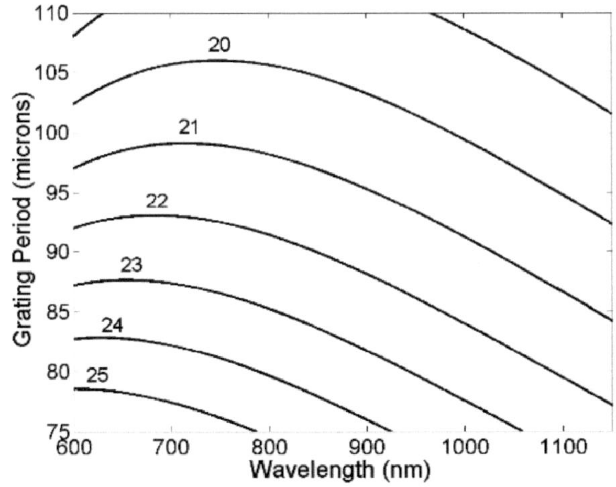

Figure 3.
Phase matching curves of the 20–25th cladding modes of an optical fibre with a cut-off wavelength of 670 nm. The relationship between the grating period and wavelength is shown. Reprinted with permission from Ref. [28], OSA.

can be observed that, at the turning point, the gradient of the curve tends to zero, $|d\Lambda/d\lambda| \rightarrow 0$ (and $|d\lambda/d\Lambda| \rightarrow \infty$). The waveguide dispersion of an LPG, expressed as $\gamma = (d\lambda/d\Lambda)\,\Delta n_{core}$ [29], will tend to infinity. As γ can be used to generalise the sensitivity of an LPG [11] it indicates that the transmission spectrum of an LPG that is fabricated with a period that closely matches a turning point, will have the highest sensitivity to external perturbations [11, 14]. The appearance of turning points will move towards shorter wavelengths when the cladding mode order is increased, as presented in **Figure 3**.

With increasing wavelength, the cladding mode's effective refractive index will decrease more than the effective refractive index of the core mode [15, 30]; this corresponds to the dual bands that become apparent in the LPG transmission spectrum.

Figure 4 shows how the grating period of an LPG approaching the turning point of the LP_{021} mode affects the transmission spectrum of the LP_{020} and LP_{021} cladding modes. **Figure 4(b)** shows the transmission spectrum where the period of the LPG chosen does not cut across the phase matching curve (**Figure 4(a)**) of LP_{021} at the turning point. A single resonant band develops (**Figure 4(d)**) followed by a small change in the central wavelength of the LP_{020} band as the period of the LPG hits the PMTP (**Figure 4(c)**). As the LPG crosses the turning point, the single band will split leading to the formation of two resonance bands. A much larger evolution can be seen for the LP_{021} mode when compared to the LP_{020} mode due to the much smaller gradient of the phase matching curve.

When an external perturbation is applied to the LPG, the two resonance bands of the mode around the turning point can either move towards or away from each other. This depends on the perturbation and the initial period of the LPG. The two bands respond differently to each other due to a non-symmetrical resonance [14], which may be due to modal dispersion [31]. For a 202.5 μm period LPG with dual resonance bands around the turning point, exposed to different temperatures, each band shows a sensitivity of 2.54 nm/°C (red shifted) and 3.29 nm/°C (blue shifted), respectively [14]. In the circumstance where the two bands shift towards each other, a single broad bandwidth band appears, similar to what is shown in **Figure 4(d)**. Under the influence of an external measurand, the coupling strength between the core mode and the cladding mode changes, altering only the amplitude of this single band and not the resonant wavelength [15].

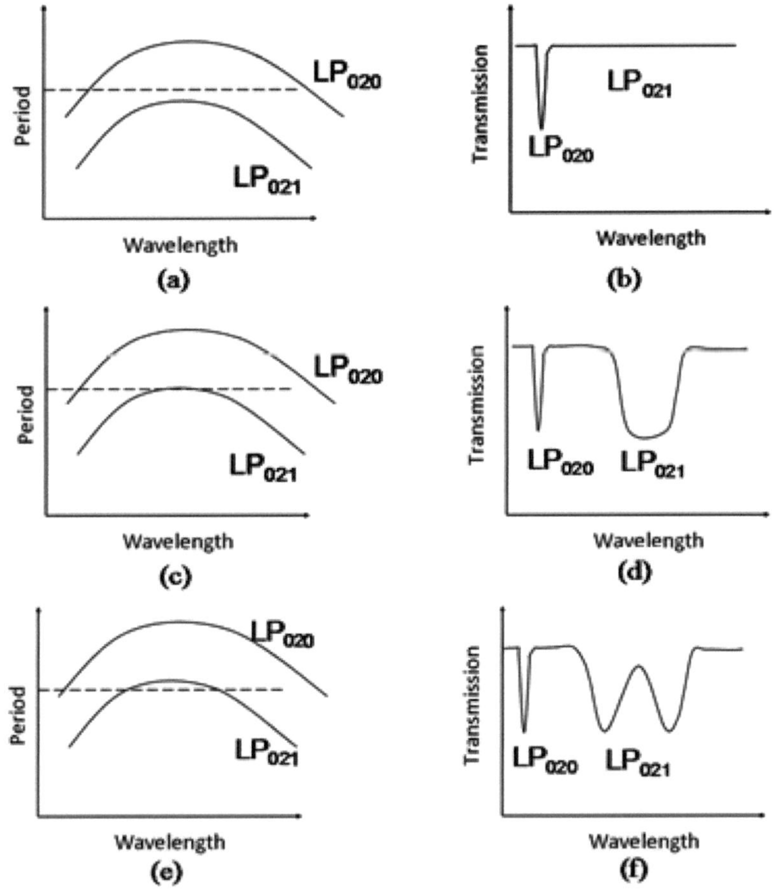

Figure 4.
Illustration of the phase matching curve ((a), (c), (e)) with their corresponding transmission spectrum ((b), (d), (f)) for cladding modes LP_{020} and LP_{021} as coupling approaches and crosses the turning point. Korposh et al., adapted from [31]; originally published under CC BY 3.0 licence. Available from: 10.5772/52935

3. Fabrication

LPGs at PMTP can be fabricated using the same methods as those used for conventional LPGs, albeit with higher precision - the effective refractive indices of the optical fibre modes can be altered via photo-induction, or by physical deformation [7]. The refractive index can be altered using a number of different methods. Some of these include local exposure of the fibre to a UV laser [28, 32, 33, 34], CO_2 laser [35, 36], femtosecond laser [23, 37] or by electrical arc discharge [30, 38]. PMTPs have been written in conventional single mode and doped fibres [12, 11, 33, 34, 36, 39], and have been theoretically investigated using photonic crystal fibres [23]. The length of an LPG tends to range from 30 to 50 mm and have refractive index changes of around 10^{-4} [40]. Coelho et al. [41] calculated a refractive index change in the order of 2×10^{-4} and 3×10^{-4} for the core and cladding, respectively when writing an LPG in a single mode fibre using mid-infrared laser radiation. The same order of magnitude of refractive index change (4.49×10^{-4}) is also needed for femtosecond laser radiation [42].

A PMTP can also be tuned after an LPG has been fabricated, by tuning the mode coupling and effective index guiding via the means of tapering [12], UV exposure [43], with a thin film overlay [31, 28], etching [44] and radiation exposure [12].

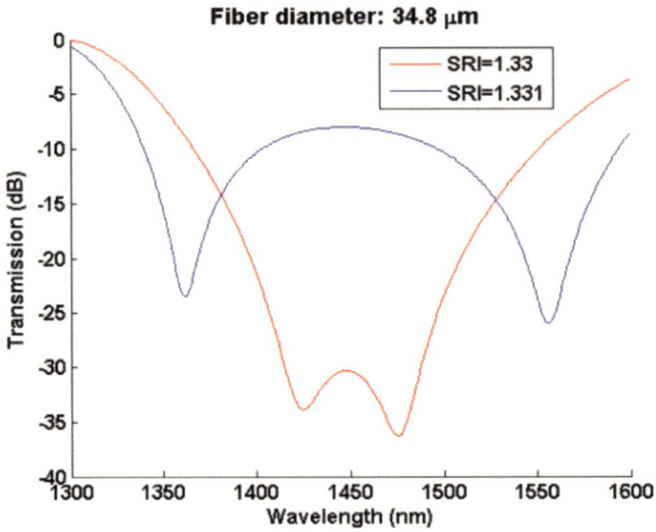

Figure 5.
Transmission spectra of a PMTP LPG with a 34.8 μm cladding diameter and 288.5 μm period, showing a large wavelength shift with a surrounding refractive index change of 0.001. SRI is surrounding refractive index. Reprinted with permission from Ref. [49], OSA.

The LPG can also been enhanced by reducing the cladding via hydrofluoric (HF) acid [13, 45–47] and plasma [48] etching to tailor the coupling strength of the cladding mode at PMTP. Using this method, Biswas et al. were able to increase the refractive index sensitivity of a hydrogen loaded PMTP LPG with a 165 μm period from 1350 to 1847 nm/RIU [45]. A refractive index sensitivity reported by Villar is 143×10^3 nm/RIU [49]. **Figure 5** shows the resonance band of the LPG splitting into two separate bands with a wavelength separation of approximately 200 nm, with a change of 0.001 RIU. This was theoretically obtained by reducing the diameter of a SMF28 single mode to 34.8 μm whilst operating at a period close to PMTP at 288.5 μm [49].

Plasma etching via ion bombardment and chemical reaction has been used to etch the fibre cladding of an LPG to bring the resonance closer to the turning point. This process assists in the precise post processing of nano-coated fibres in hard and chemically resistant films, for example, diamond-like carbon [50]. Radiation exposure has also been demonstrated to alter the refractive index of B-Ge co-doped fibres, with an equivalent increase in core refractive index of around 1×10^{-5} [12].

3.1 Fabrication considerations

Due to the nature of the LPG, they can be highly sensitive to the surrounding environment. There are stringent demands placed on the fabrication process and the system used in order to fabricate LPGs at PMTP reproducibly. The notable constraints are given by ambient temperature (**Figure 6**), duty cycle [32], power of the irradiation source [35] and amplitude of the index modulation [51]. The difference in the final outcome of LPG spectra where the ambient temperature is not controlled and allowed to fluctuate, and maintained to ±0.5°C are shown in **Figure 6**(**a**) and (**b**), respectively.

A period change of less than 1 μm can also influence transmission spectrum significantly and high resolution control has to be taken into account when deciding on the grating period [32, 35]. UV exposure time may also play a part in the sensitivity of LPGs at turning point; the spectrum of a 168.7 μm period PMTP LPG written in

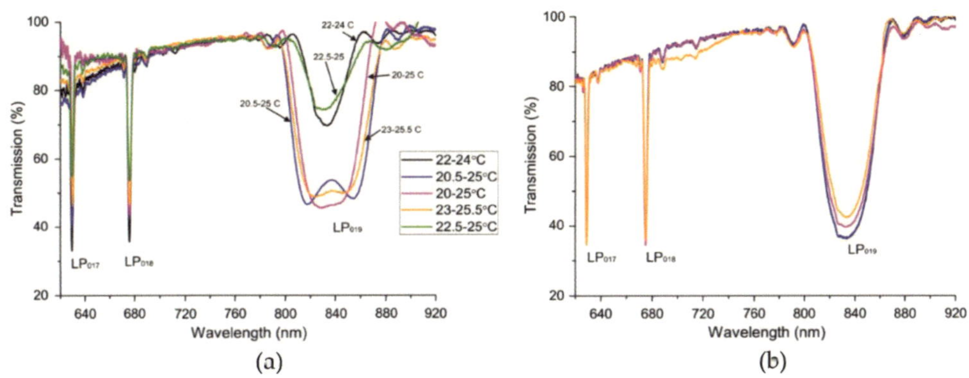

Figure 6.
Transmission spectra of a 110.9 µm PMTP LPG. The temperature is (a) not controlled (5 spectra) and (b) controlled to ±0.5°C (4 spectra). Reprinted with permission from Ref. [32], OSA.

boron co-doped fibre had a greater variation with pressure when fabricated with a longer exposure time [34]. Other factors that can affect the grating include the size of the fibre. By changing the diameter of the cladding, but maintaining the same period, the dual resonance bands will also change accordingly [44].

Hydrogen loading can induce or increase the photosensitivity in a fibre by increasing the effective refractive index difference between the core and cladding [52]. However, hydrogen will diffuse from the fibre gradually over time, causing the LPG spectrum to drift [52, 53]. Annealing a hydrogen loaded fibre at a temperature above the desired operating temperature can help overcome this problem [54]. This rapid removal of hydrogen will still cause the resonance wavelengths to shift, due to the changing effective indices, but will remain stable and permanent after the annealing process has been completed. This has to be taken into consideration when choosing a period to fabricate an LPG, at or around turning point, using a hydrogen loaded fibre [33, 43].

4. Applications

For an LPG to function at its optimum sensitivity when exposed to an external perturbation, its period should be chosen such that it is able to operate at a turning point. Optical LPGs operating at the turning point provide the potential for low cost sensors with fast response time [21, 55, 56] and can provide a simpler detection method as some are able to work as intensity-based sensors [15, 17, 55].

LPGs operating at the PMTP have been used for temperature, strain, refractive index sensing [11, 35] and as filters. PMTP LPGs, when modified with a functional film can be adapted for potential uses as enhanced gas and chemical sensors [28].

4.1 Filters

By employing the broadband characteristics, PMTP LPGs can make successful bandpass and rejection filters [57, 58]. A coated PMTP LPG with a π phase shift is simulated to provide tuneable broadband characteristics for rejection filtering applications [58]. By introducing multiple π phase shifts, it is possible to adjust the separation between the dual resonant bands. On the other hand, by partially coating a phase shifted PMTP LPG, bandgaps appear over a narrow wavelength band which could be useful for designing spectral filters [59].

4.2 Temperature sensing

By careful choice of the grating period to allow coupling close to or at the turning point, it is possible to improve the temperature sensitivity of an LPG [14, 15, 33, 60]. Shu et al. [14] showed that for an LPG with a 175 μm period, the dual resonance band had a temperature sensitivity of 3.2 nm/°C whereas the single band away from the PMTP had a much lower sensitivity of -0.31 nm/°C. The response of the dual resonant bands to changing temperature is non-linear, with a reduction in the rate of separation as the resonance moves away from the phase matching turning point. A band operating away from the turning point has a linear response to changing temperatures [60], which may make it easier to characterise temperature sensitivity. When comparing the sensitivities of LPGs with periods of 110.8, 111 and 111.5 μm the highest sensitivity was seen when the LPG was chosen to operate near the PMTP (111.5 μm period), just as the single broad resonance band would begin to appear. A sensitivity of 0.99 nm/°C was achieved for the sensor at turning point, which was more than five times greater when compared to the sensitivity of a band away from turning point (0.17 nm/°C). As the temperature response changes depending on the surrounding environment, it may also be possible for the thermo-optic coefficients of a surrounding medium to be characterised [60].

4.3 Strain sensing

Previous studies have shown that the dual resonance bands will move together, when increasing strain is applied, with a near linear trend [14]. The separation of the dual bands was calculated to be -33.6 nm/1000 μm whereas the sensitivity of an LPG can be more than an order of magnitude less [16]. Using the single broad band resonant mode at PMTP, Grubsky et al. [15] were able to obtain a sensor resolution of 1 μm by changing the coupling strength using different strengths of strain; the band would show an appreciable decrease in amplitude as strain increased, whilst the wavelength remained fixed as shown in **Figure 7**. The fixed wavelength allows for a simpler detection method as spectrometers or post processing can be bypassed for a simple photodetector [15].

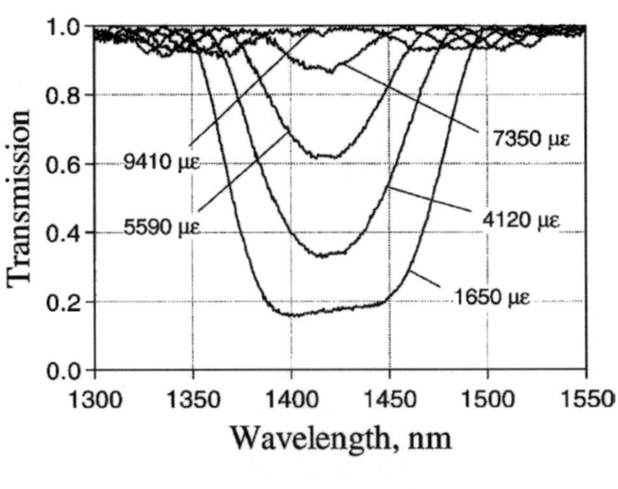

Figure 7.
Transmission spectrum of a 50.1 μm period LPG with increasing strain. The wavelength of the band remains fixed at 1420 nm but the coupling efficiency decreases. Reprinted with permission from Ref. [15], OSA.

4.4 Refractive index sensing

The optical intensities of the guided cladding modes will likely dissipate out of the fibre after a short distance. This could be due the fibre being bent, of from scattering or absorption due to the protective jacket of the fibre. By removing the jacket, the cladding mode evanescent field extends into the immediate surroundings, which will influence the fibre mode properties. The refractive index of the local environment will affect the effective refractive indices of the cladding modes propagating in the fibre. These effective refractive indices determine the sensitivity of the PMTP LPG.

A refractive index PMTP LPG sensor based on intensity as opposed to wavelength shift provides high sensitivity with a linear response for different refractive indices. A linear correlation coefficient, of more than 0.98 and sensitivity of 59.88/RIU for the refractive index range of 1.410–1.420 was achieved using a PMTP LPG with a period of 231.5 μm [17]. This allows for simple calibration and linear interpolation to determine the sensitivity of the sensor within this refractive index range.

By coating a PMTP LPG, such that it coincides with the mode transition region, it can be possible to enhance the refractive index sensitivity of a sensor [18, 31]. Mode transition describes the reorganisation of cladding modes caused when a higher refractive index material of a certain optical thickness surrounds the LPG [61–63]. After a certain thickness, the surrounding material is able to guide the outer most cladding mode, causing large shifts in the resonance wavelengths. The sensitivity of the LPG can also be optimised by controlling the optical thickness of the overlay so that the turning point coincides with the mode transition region, and has been proven theoretically and experimentally [28, 64]. Pilla et al. [18] were able to achieve a sensitivity exceeding 9000 nm/RIU in solutions with RIs similar to water, using an LPG with a single resonance band close to 1.55 μm. By increasing the refractive index, the dual bands appear and eventually split as shown in **Figure 8**. A mesoporous coating consisting of silica nanospheres was able to improve the refractive index sensitivity of a 100 μm period LPG operating near turning point, with a maximum sensitivity of 1927 ± 59 nm/RIU, as well as increase the detection range of the LPG [19]. The refractive index sensitivity of the first electric arc induced LPG at turning point was increased from 400 to 700 nm/RIU to 887–2146 nm/RIU

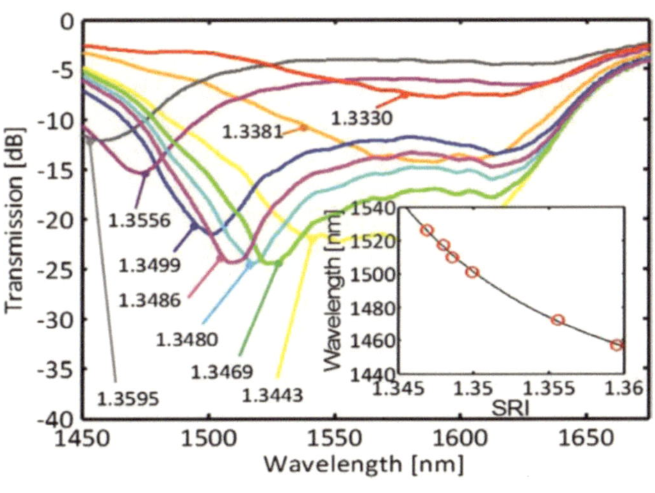

Figure 8.
Transmission spectra showing the response to different refractive indices of ethanol solutions. SRI is surrounding refractive index. Reprinted with permission from Ref. [18], OSA.

by operating close to the turning point and by coating a thin film overlay of silicon nitride [20]. By combining these phenomena with a reduced diameter fibre, it can be possible to further enhance the sensitivity of the LPG, which can help improve the resolution of biochemical sensing applications.

4.5 Chemical and gas sensing

Their small dimensions, suitability in harsh environments and versatility make fibres ideal sensing platforms. The high sensitivity property of the PMTP allows for detection of low quantities and concentrations of different chemicals. For example, a PMTP is able to detect a 0.01% aqueous solution of cane sugar [65].

Optical fibre sensors with functional thin film coatings have become of interest due to the large pool of possible applications, especially in the chemical and biosensing fields. These sensors have the potential to measure concentrations of chemicals or for detecting gaseous species. These thin films can improve the sensing ability of the fibre and allow them to have different responses to different stimuli, such as concentrations of chemicals in the surrounding environment [19, 66, 67]. Functional materials can also be used to enhance the sensitivity for detection of a particular analyte. The thickness of the film on the LPG sensor is usually in the region of a few 100 nm as the transmission spectrum can be greatly affected [28, 61, 68, 69].

The following techniques allow for nanoscale thickness deposition control of the coating. These include the Langmuir–Blodgett deposition [28, 59, 70], self-assembly [21, 55, 71], layer-by-layer deposition [19], atomic layer deposition [72], sol-gel [73] and liquid phase deposition [67].

Functionalised LPGs operating around the PMTP have been tested for volatile organic compound (VOC) detection [70, 74, 75]. VOCs can be generated from a variety of processes. These include, but are not limited to, fuels and combustion processes, petroleum products, paints, and in nature and farming [76]. PMTP LPGs have been used for toluene [70, 74] and benzene [74] detection. By applying a functional overlay, particular compounds will affect the refractive index of the overlay and therefore influence the fibre modes, which will be shown as changes in transmission spectrum [74]. Providing clean water is an integral part of life, therefore monitoring water quality is critical. Partridge et al. demonstrated a proof-of-concept sensor for detecting toluene contamination in water. A 97 μm period LPG at turning point coated with calix [4]res C11 and was shown to be specific to toluene when compared to another potential contaminant, ethanol as shown in **Figure 9(a)**. The sensor was able to achieve a minimum detection limit of 100 ppm (see **Figure 9(b)**) which is the approximate limit of oil weep sampling and leaking oil plumes [70]. Some gases, such as hydrogen, are odourless and colourless, and have a low ignition energy. Means of detecting leaks in small quantities are therefore an important safety tool. A sensor coated with a 70 nm thick palladium overlay, when exposed to 4% hydrogen, experienced a dual band wavelength shift apart of 7.5 nm [77]. A thin film PMTP LPG with a functional material of poly(acrylic acid) PAA was successfully used to selectively bind to ammonia with lower detection levels when compared to other devices such as colorimetric and absorption spectroscopic devices [71].

Optical sensors have become more popular and valuable in the biomedical field. They have the potential to be used for diagnosis and monitoring and can be cost effective, portable and easy to use. This has also contributed to the increase in interest in label-free sensing using LPGs, especially at the PMTP where there is high RI sensitivity, rapid response and adaptability by choice of overlay [18, 72, 78, 79]. LPGs at PMTP have been used for real time monitoring of phage-bacteria interactions [80, 81] where a 1.3 nm wavelength shift was detected as bacteria binding

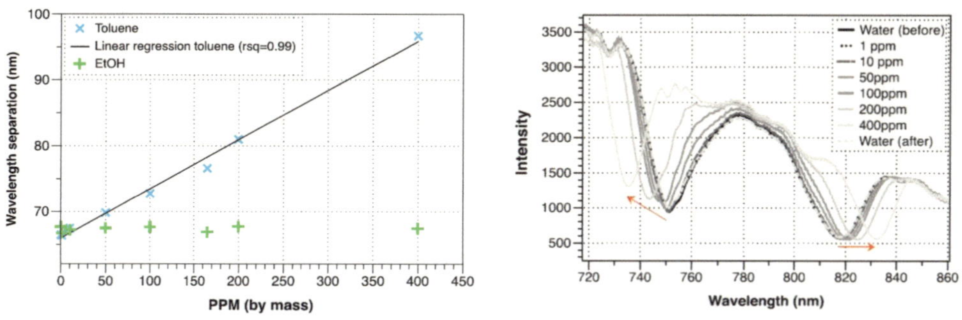

(a) (b)

Figure 9.
(a) Plots showing the response of a calix [4]red C11 coated LPG sensor to toluene and ethanol concentrations. Exposure to ethanol shows negligible response compared to toluene. (b) Transmission spectra showing the response of the dual resonance bands of a calix [4]red C11 coated LPG to different concentrations of toluene. Partridge et al.,. Reprinted from [70]; originally published under CC BY 3.0 licence. Available from: 10.1016/j.snb.2014.06.121

occurred [80], and target-probe DNA hybridisation [82, 83]. The well-known properties of the streptavidin-biotin interaction, as well as its use for studying biological processes, have encouraged its use as a means of understanding the characteristics of thin film PMTP sensors [21, 55]. Korposh et al. showed a mesoporous SiO_2 film coated sensor with an additional functional material could detect a specific chemical species. In this case the species was a porphyrin compound, with a 10 μM concentration being detected in under 10s [56].

Selectivity is an important indicator for a sensor as it can potentially prevent false readings, which is especially helpful at the highly sensitive turning point region [66, 67, 84]. Molecular imprinting provides a versatile platform as the properties of the receptor can be modified to detect a desired molecular compound [66, 67]. An LPG coated with a molecularly imprinted polymer (MIP) was prepared, for the detection of antibiotics [66]. In the presence of different commonly prescribed antibiotics, the sensor showed selectivity to the target antibiotic vancomycin. The target compound can also be removed and the sensor reused. Removal methods used include organic solvents, and photodecomposition have also been investigated [67]. Reusing an LPG or sensor also increases the versatility of a biosensor [81, 82, 84] and can therefore be more time and cost effective.

Choosing a particular fibre type can also contribute to the final characteristics of the fabricated sensor. For instance, PMTP LPGs written in boron co-doped fibres have been demonstrated as radiation dose sensors [36], pressure sensors [34] (boron co-doping can increase the pressure-optic coefficient of a material [85]), and for fuel adulteration detection [39].

4.6 Sensor limitations

The inherent high sensitivity of the PMTP LPG also leads to its limitations. For instance, cross-talk or unwanted interference, such as from varying temperature, have to be limited in order to ensure the shift in the wavelengths is only due to the desired parameter. By fabricating cascaded PMTP LPGs, based on a Mach-Zehnder interferometer, it is possible to eliminate interference [86] and make simultaneous measurements for parameter compensated sensing [87]. James et al. demonstrated that coating a cascaded PMTP LPG device with mesoporous silica nanoparticles, and subsequently infusing a functional material to the central of the region (between the two gratings) enables measuring of only the desired analyte [86].

The broad spectral width of the resonance bands can also limit the multiplexing capabilities of the PMTP LPG. By utilising the double resonance bands, simultaneous measurements of surrounding refractive index and temperature were carried out for temperature ranges limited to ±3°C if the refractive index range is ±0.004 RIU [87]. This information enables temperature calibrated sensing.

5. Summary

PMTP LPGs as versatile sensing platforms have become increasingly popular. The ultra-high sensitivity and quick response, as compared to other configurations of LPGs, proposes promising capabilities for use in many sensing applications. PMTP LPGs can be achieved by the precise choice of period or by post processing methods. However, the nature of their high sensitivity can also pose as limitations and sensors need to be optimised in order to avoid interference. By applying functional nanoscale coatings, it has been possible to tailor PMTP LPGs to have a preferred sensitivity to particular parameters. This opens the doors for a number of applications in the medical field for portable, real time monitoring. The sensors have potential deployment into the biochemical industry for measuring chemical concentrations and can also be applied to sense different gases in the environment. Specially designed sensors can also be used for monitoring food quality. However, calibration must be carried out first as the period and thickness of coating will affect the sensitivity of the sensor. More vigorous and consistent testing needs to be carried out before adoption in the healthcare and food safety industry [2]. Sensor packaging, ease of use and reusability are some aspects that need to be taken into consideration. As many chemical sensing applications take place in a solution, it may be beneficial to fabricate an LPG such that the appearance of the dual resonance bands will appear when the fibre is placed in solution. The fibre cladding size can greatly affect the sensitivity allowing a greater flexibility when designing an LPG at PMTP [49].

Author details

Rebecca Yen-Ni Wong[1*], Dora Hu Juan Juan[1], Morten Ibsen[2] and Perry Ping Shum[3,4]

1 Institute for Infocomm Research, Agency for Science, Technology and Research (A*STAR), Singapore

2 Optoelectronics Research Centre, University of Southampton, Southampton, United Kingdom

3 Centre for Optical Fibre Technology, School of Electrical and Electronic Engineering, Nanyang Technological University, Singapore

4 CINTRA CNRS/NTU/THALES, UMI 3288, Research Techno Plaza, Singapore

*Address all correspondence to: rebecca-wong@i2r.a-star.edu.sg

References

[1] Alwis L, Sun T, Grattan KTV. Developments in optical fibre sensors for industrial applications. Optics and Laser Technology. 2016;**78**(A): 62-66. DOI: 10.1016/j.optlastec. 2015.09.004

[2] Correia R, James S, Lee SW, Morgan SP, Korposh S. Biomedical application of optical fibre sensors. Journal of Optics. 2018;**20**(7):1-25. DOI: 10.1088/2040-8986/aac68d

[3] Ramakrishnan M, Rajan G, Semenova Y, Farrell G. Overview of Fiber optic sensor Technologies for Strain/temperature sensing applications in composite materials. Sensors. 2016;**16**(99):1-27. DOI: 10.3390/ s16010099

[4] Mescia L, Prudenzano F. Advances on optical Fiber sensors. Fibers. 2014;**2**(1):1-23. DOI: 10.3390/ fib2010001

[5] Vengsarkar AM, Lemaire PJ, Judkins JB, Bhatia V, Erdogan T, Sipe JE. Long-period fiber gratings as band-rejection filters. Journal of Lightwave Technology. 1996;**14**(1):58-65. DOI: 10.1109/50.476137

[6] Vengsarkar AM, Lemaire PJ, Judkins JB, Bhatia B, Erdogan T, Sipe JE. Long-period fiber gratings as band-rejection filters. In: Optical Fiber Communications PD4; San Diego; 1995

[7] James SW, Tatam RP. Optical fibre long-period grating sensors: Characteristics and application. Measurement Science and Technology. 2003;**14**(5):R49-R61. DOI: 10.1088/0957-0233/14/5/201

[8] Kashyap R. Fibre Bragg Gratings. 2nd ed. New York: Academic Press; 2010. DOI: 10.1016/C2009-0-16830-7

[9] Werneck MMM, Allil RCSB, de Nazaré FVB. Fiber Bragg Gratings: Theory, Fabrication, and Applications. Washington: SPIE; 2017. DOI: 10.1117/3.2286558

[10] Liu Z, Tam HY. Industrial and medical applications of fiber Bragg gratings (invited paper). Chinese Optics Letters. 2016;**14**(12):120007, 10.3788/ COL201614.120007

[11] Shu X, Zhang L, Bennion I. Sensitivity characteristics near the dispersion turning points of long-period fiber gratings in B/Ge codoped fiber. Optics Letters. 2001;**26**(22):1755-1757. DOI: 10.1364/OL.26.001755

[12] Chaubey S, Kher S, Oak SM. Radiation and taper tuning of long period grating for high sensitivity strain measurement. In: 7th Workshop on Fibre and Optical Passive Components. Montreal: Proceedings of the IEEE; 2011

[13] Chen X, Zhou K, Zhang L, Bennion I. Dual-peak long-period fiber gratings with enhanced refractive index sensitivity by finely tailored mode dispersion that uses the light cladding etching technique. Applied Optics. 2007;**46**(4):451-455. DOI: 10.1364/AO.46.000451

[14] Shu X, Shang L, Bennion I. Sensitivity characteristics of long-period fiber gratings. Journal of Lightwave Technology. 2002;**20**(2): 255-266. DOI: 10.1109/50.983240

[15] Grubsky V, Feinberg J. Long-period fiber gratings with variable coupling for real-time sensing applications. Optics Letters. 2000;**25**(4):203-205. DOI: 10.1364/OL.25.000203

[16] Bhatia V, Campbell DK, Sherr D, D'Alberto TG, Zaboronick NA, Eyck GAT, et al. Temperature-insensitive and strain-insensitive long-period grating

sensors for smart structures. Optical Engineering. 1997;**36**(7):1872-1876. DOI: 10.1117/1.601379

[17] Lan X, Han Q Huang J, Wang H, Gao Z, Kaur A, et al. Turn-around point long-period fiber grating fabricated by CO_2 laser for refractive index sensing. Sensors and Actuators B: Chemical. 2013;**177**:1149-1155. DOI: 10.1016/j.snb.2012.12.006

[18] Pilla P, Trono C, Baldini F, Chiavaioli F, Giordano M, Cusano A. Giant sensitivity of long period gratings in transition mode near the dispersion turning point: An integrated design approach. Optics Letters. 2012;**37**(19):4152-4154. DOI: 10.1364/OL.37.004152

[19] Korposh S, Lee SW, James SW, Tatam RP. Refractive index sensitivity of fibre-optic long period gratings coated with SiO_2 nanoparticle mesoporous thin films. Measurement Science and Technology. 2011;**22**(7):1-10. DOI: 10.1088/0957-0233/22/7/075208

[20] Debowska AK, Smietana M, Mikulic P, Bock WJ. High temperature nano-coated electric-arc-induced long-period gratings working at the dispersion turning point for refractive index sensing. Japanese Journal of Applied Physics. 2014;**53**(8S2)::08ME01-1-08ME01-5). DOI: 10.7567/JJAP.53.08ME01

[21] Marques L, Hernandez FU, James SW, Morgan SP, Clark M, Tatam RP, et al. Highly sensitive optical fibre long period grating biosensor anchored with silica core gold shell nanoparticles. Biosensors and Bioelectronics. 2016;**75**:222-231. DOI: 10.1016/j.bios.2015.08.046

[22] Ghambir M, Gupta S. Review of turn around point long period fiber gratings. Journal of Sensor Technology. 2015;**5**(4):81-89. DOI: 10.4236/jst.2015.54009

[23] Kanka J. Design of turn-around-point long-period gratings in a photonic crystal fiber for refractometry of gases. Sensors and Actuators B: Chemical. 2013;**182**:16-24. DOI: 10.1016/j.snb.2013.02.048

[24] Erdogan T. Cladding-mode resonances in short- and long- period fiber grating filters. Journal of the Optical Society of America. A. 1997;**14**(8):1760-1773. DOI: 10.1364/JOSAA.14.001760

[25] Erdogan T. Cladding-mode resonance in short- and long- period fiber grating filters: Errata. Journal of the Optical Society of America. A. 2000;**17**(11):2113. DOI: 10.1364/JOSAA.17.002113

[26] Huang WP. Coupled-mode theory for optical waveguides: An overview. Journal of the Optical Society of America. A. 1994;**11**(3):963-983. DOI: 10.1364/JOSAA.11.000963

[27] Gloge D. Weakly guiding fibers. Applied Optics. 1971;**10**(10):2252-2258. DOI: 10.1364/AO.10.002252

[28] Cheung CS, Topliss SM, James SW, Tatam RP. Response of fibre optic long period gratings near the phase matching turning point to the deposition of nanostructured coatings. Journal of the Optical Society of America B: Optical Physics. 2008;**25**(6):897-902. DOI: 10.1364/JOSAB.25.000897

[29] MacDougall TW, Pilevar S, Haggans CW, Jackson MA. Generalized expression for the growth of long period gratings. IEEE Photonics Technology Letters. 1998;**10**(10):1449-1451. DOI: 10.1109/68.720290

[30] Rego G. Arc-induced long period fiber gratings. Journal of Sensors. 2016;**2016**:1-13. DOI: 10.1155/2016/3598634

[31] Korposh S, James S, Tatam R, Lee SW. Optical fibre long-period gratings functionalised with nano-assembled thin-films: Approaches to chemical sensing. In: Cuadrado-Laborde C, editor. Current Trends in Short- and Long-Period Fiber Gratings. Rijeka: InTech; 2013. pp. 63-86. DOI: 10.5772/52935

[32] Wong RYN, Chehura E, Staines SE, James SW, Tatam RP. Fabrication of fiber optic long period gratings operating at the phase matching turning point using an ultraviolet laser. Applied Optics. 2014;**53**(21):4669-4674. DOI: 10.1364/AO.53.004669

[33] Mizunami T, Fukuda T, Hayashi A. Fabrication and characterization of long-period-grating temperature sensors using Ge-B-co-doped photosensitive fibre and single-mode fibre. Measurement Science and Technology. 2004;**15**(8):1467-1473. DOI: 10.1088/0957-0233/15/8/006

[34] Smietana M, Bock WJ, Mikulic P, Chen J. Tuned pressure sensitivity of dual resonant long-period gratings written in boron co-doped optical fiber. Journal of Lightwave Technology. 2012;**30**(8):1080-1084. DOI: 10.1109/JLT.2011.2169234

[35] Lan X, Han Q Wei T, Huang J, Xiao H. Turn-around-point long-period fiber gratings fabricated by CO_2 laser point-by-point irradiation. IEEE Photonics Technology Letters. 2011;**23**(22):1664-1666. DOI: 10.1109/LPT.2011.2166256

[36] Kher S, Chaubey S, Kashyap R, Oak SM. Turnaround-point long-period fiber gratings (TAP-LPGs) as high-radiation-dose sensors. IEEE Photonics Technology Letters. 2012;**24**(9):742-744. DOI: 10.1109/LPT.2012.2187637

[37] Shen F, Zhou K, Zhang L, Shu X. Long period Fiber grating around the dispersion turning point fabricated with a femtosecond laser. In: Proceedigs of OSA. Guangzhou: ACPC; 2017

[38] Colaço C, Caldas P, Del Villar I, Chibante R, Rego G. Arc-induced long-period Fiber gratings in the dispersion turning points. Journal of Lightwave Technology. 2016;**34**(19):4584-4590. DOI: 10.1109/JLT.2016.2540678

[39] Kher S, Chaubey S, Kishore J, Oak SM. Detection of fuel adulteration with high sensitivity using turnaround point long period fiber gratings in B/Ge doped fibers. IEEE Sensors Journal. 2013;**13**(11):4482-4486. DOI: 10.1109/JSEN.2013.2270312

[40] Martinez-Rios A, Monzon-Hernandez D, Torres-Gomez I, Salceda-Delgado G. Long period fibre gratings. In: Yasin M, editor. Fiber Optic Sensors. Rijeka: InTech; 2012. pp. 275-294. DOI: 10.5772/27727

[41] Coelho JMP, Nespereira MC, Abreu M, Rebordao JM. Modeling refractive index change in writing long-period fiber gratings using mid-infrared laser radiation. Photonic Sensors. 2013;**3**(1):67-73. DOI: 10.1007/s13320-012-0084-1

[42] Ahmed F, Joe HE, Min BK, Jun MBG. Characterization of refractive index change and fabrication of long period gratings in pure silica fiber by femtosecond laser radiation. Optics and Laser Technology. 2015;**74**:119-124. DOI: 10.1016/j.optlastec.2015.05.018

[43] Shu X, Zhu X, Wang Q Jiang S, Shi W, Huang Z, et al. Dual resonant peaks of LP015 cladding mode in long-period gratings. Electronics Letters. 1999;**35**(8):649-651. DOI: 10.1049/el:19990442

[44] Ling Q Gu Z, Gao K. Smart design of a long-period fiber grating refractive index sensor based on dual-peak resonance near the phase-matching turning point. Applied Optics. 2018;**57**(10):2693-2697. DOI: 10.1364/AO.57.002693

[45] Biswas P, Basumallick N, Bandyopadhyay S, Dasgupta K, Ghosh A, Bandyopadhyay S. Sensitivity enhancement of turn-around-point long period gratings by tuning initial coupling condition. IEEE Sensors Journal. 2015;**15**(2):1240-1245. DOI: 10.1109/JSEN.2014.2361166

[46] Mysliwiec M, Grochowski J, Krogulski K, Mikulic P, Bock WJ, Smietana M. Effect of wet etching of arc-induced long-period gratings on their refractive index sensitivity. Acta Physica Polonica, A. 2013;**124**(3): 521-524. DOI: 10.12693/APhysPolA.124.521

[47] Szymanska M, Krogulski K, Mikulic P, Bock WJ, Smietana M. Sensitivity of long-period gratings modified by their bending. In: Proceedia Engineering, 28th Eurosensors; Brescia; 2014. pp. 1180-1183

[48] Smietana M, Koba M, Mikulic P, Bock WJ. Measurements of reactive ion etching process effect using long-period fiber gratings. Optics Express. 2014;**22**(5):5986-5994. DOI: 10.1364/OE.22.005986

[49] Del Villar I. Ultrahigh-sensitivity sensors based on thin-film coated long period gratings with reduced diameter, in transition mode near the dispersion turning point. Optics Express. 2015;**23**(7):8389-8398. DOI: 10.1364/OE.23.008389

[50] Smietana M, Koba M, Mikulic P, Bock WJ. Tuning properties of long-period gratings by plasma post-processing of their diamond-like carbon nano-overlays. Measurement Science and Technology. 2014;**25**:1-7. DOI: 10.1088/0957-0233/25/11/114001

[51] Partridge M, James S, Barrington J, Tatam R. Overwrite fabrication and tuning of long period gratings. Optics Express. 2016;**24**(20):22345-22356. DOI: 10.1364/OE.24.022345

[52] Mizunami T, Fukuda T. FEM calculation and the effects of hydrogen diffusion in fabrication processes of long-period fiber gratings. Optics Communication. 2006;**259**(2):581-586. DOI: 10.1016/j.optcom.2005.09.002

[53] Namihira Y, Mochizuki K, Kuwazuru M, Iwamoto Y. Effects of hydrogen diffusion on optical fibre loss increase. Electronics Letters. 1983;**19**(24):1034-1035. DOI: 10.1049/el:19830701

[54] Libish TM, Bobby MC, Linesh J, Mathew S, Biswas P, Bandyopadhyay S, et al. The effect of annealing and temperature on transmission spectra of long period gratings written in hydrogen loaded standard single mode fiber. Optik. 2013;**124**(20):4345-4348. DOI: 10.1016/j.ijleo.2013.02.010

[55] Wang Z, Helfin JR, Cott KV, Stolen RH, Ramachandran S, Ghalmi S. Biosensors employing ionic self-assembled multilayers adsorbed on long-period fiber gratings. Sensors and Actuators B: Chemical. 2009;**139**(2):618-623. DOI: 10.1016/j.snb.2009.02.073

[56] Korposh S, James SW, Lee SW, Topliss S, Cheung SC, Batty WJ, et al. Fiber optic long period grating sensors with a nanoassembled mesoporous film of SiO_2 nanoparticles. Optics Express. 2010;**18**(12):13227-13228. DOI: 10.1364/OE.18.013227

[57] Ramachandran S, Ghalmi S, Wang Z, Yan M. Band-selection filters with concatenated long-period gratings in few-mode fibers. Optics Letters. 2002;**27**(19):1678-1680. DOI: 10.1364/OL.27.001678

[58] Chen H, Gu Z. Filtering characteristics of film-coated long-period fiber gratings operating at the phase-matching turning point. Optik. 2014;**125**(20):6003-6009. DOI: 10.1016/j.ijleo.2014.07.060

[59] James SW, Topliss SM, Tatam RP. Properties of length-apodized phase-shifted LPGs operating at the phase matching turning point. Journal of Lightwave Technology. 2012;**30**(13):2203-2209. DOI: 10.1109/JLT.2012.2195473

[60] Korposh S, Wong R, James S, Tatam R. Temperature and thermo-optic coefficient measurements using optical fibre long period gratings operating at phase matching turning point. In: Proceedings of SPIE, 5th EWOFS; Krakow; 2013

[61] Del Villar I, Achaerandio M, Matias IR, Arregui FJ. Deposition of overlays by electrostatic self-assembly in long-period fiber gratings. Optics Letters. 2005;**30**(7):720-722. DOI: 10.1364/OL.30.000720

[62] Cusano A, Iadicicco A, Pilla P, Contessa L, Campopiano S, Cutolo A. Mode transition in high refractive index coated long period gratings. Optics Express. 2006;**14**(1):20-34. DOI: 10.1364/OPEX.14.000019

[63] Cusano A, Iadicicco A, Pilla P, Contessa L, Campopiano S, Cutolo A. Cladding mode reorganization in high-refractive index-coated long-period gratings: Effects on the refractive-index sensitivity. Optics Letters. 2005;**30**(19):2536-2538. DOI: 10.1364/OL.30.002536

[64] Brabant D, Koba M, Smietana M, Bock WJ. Analysis of mode transitions in a long-period fiber gratings with nano-overlay of diamond-like carbon. In: Proceedings of SPIE, Photonics North; Montreal; 2014

[65] Shu X, Huang D. Highly sensitive chemical sensor based on the measurement of the separation of dual resonant peaks in a 100-um-period fiber grating. Optics Communication. 1999;**171**(1-3):65-69. DOI: 10.1016/S0030-4018(99)00522-2

[66] Korposh S, Chianella I, Guerreiro A, Caygill S, Piletsky S, James SW, et al. Selective vancomycin detection using optical fibre long period gratings functionalised with molecularly imprinted polymer nanoparticles. The Analyst. 2014;**139**(9):2229-2236. DOI: 10.1039/c3an02126b

[67] Wong R, Korposh S, Lee SW, James SW, Tatam RP. Photodecomposition of a target compound detected using an optical fibre long period grating coated with a molecularly imprinted titania thin film. In: Proceedings of SPIE, 24th OFS; Curitiba; 2015

[68] Rees ND, James SW, Tatam RP, Ashwell GJ. Optical fiber long-period gratings with Langmuir-Blodgett thin-film overlays. Optics Letters. 2002;**27**(9):686-688. DOI: 10.1364/OL.27.000686

[69] Bandyopadhyay S, Biswas P, Chiavaioli F, DT K, Basumallick N, Trono C, et al. Long-period fiber grating: A specific design for biosensing applications. Applied Optics. 2017;**56**(35):9846-9853. DOI: 10.1364/AO.56.009846

[70] Partridge M, Wong R, James SW, Davis F, Higson SPJ, Tatam RP. Long period grating based toluene sensor for use with water contamination. Sensors and Actuators B: Chemical. 2014;**203**:621-625. DOI: 10.1016/j.snb.2014.06.121

[71] Wang T, Korposh S, Wong R, James S, Tatam R, Lee SW. A novel ammonia gas sensor using nanoassembled polyelectrolyte thin film on fiber-optic long-period gratings. Chemistry Letters. 2012;**41**(10):1297-1299. DOI: 10.1246/cl.2012.1297

[72] Smietana M, Koba M, Brzozowska E, Krogulski K, Nakonieczny J, Wachnicki L, et al. Label-free sensitivity of long-period gratings enhanced by atomic layer deposited

TiO_2 nano-overlays. Optics Express. 2015;**23**(7):8841-8453. DOI: 10.1364/OE.23.008441

[73] Chiavaioli F, Biswas P, Trono C, Jana S, Bandyopadhyay S, Basumallick N, et al. Sol-gel-based titania-silica thin film overlay for long period fiber grating-based biosensors. Analytical Chemistry. 2015;**87**(24):12024-12031. DOI: 10.1021/acs.analchem.5b01841

[74] Topliss SM, James SW, Davis F, Higson SPJ, Tatam RP. Optical fibre long period grating based selective vapour sensing of volatile organic compounds. Sensors and Actuators B: Chemical. 2010;**143**(2):629-634. DOI: 10.1016/j.snb.2009.10.008

[75] Korposh S, David F, James SW, Wang T, Lee SW, Higson S, et al. Detection of volatile organic compounds (VOCs) using an optical fibre long period grating with a calixarene anchored mesoporous thin film. In: Proceedings of SPIE, 5th EWOFS; Krakow; 2013

[76] Viricelle JP, Pauly A, Mazat L, Brunet J, Bouvet M, Varenne C, et al. Selectivity improvement of semi-conducting gas sensors by selective filter for atmospheric pollutants detection. Materials Science and Engineering: C. 2006;**26**(2-3):186-195. DOI: 10.1016/j.msec.2005.10.062

[77] Kim YH, Kim MJ, Rho BS, Park MS, Jang JH, Lee BH. Ultra sensitive fiber-optic hydrogen sensor based on high order cladding mode. IEEE Sensors Journal. 2011;**11**(6):1423-1426. DOI: 10.1109/JSEN.2010.2092423

[78] Pilla P, Manzillo PF, Malachovska V, Buosciolo A, Campopiano S, Cutolo A, et al. Long period grating working in transition mode as promising technological platform for label-free biosensing. Optics Express. 2009;**17**(22):20039-20050. DOI: 10.1364/OE.17.020039

[79] Chiavaioli F, Biswas P, Trono C, Bandyopadhyay S, Giannetti A, Tombelli S, et al. Towards sensitive label-free immunosensing be means of turn-around point long period fiber gratings. Biosensors and Bioelectronics. 2014;**60**:305-310. DOI: 10.1016/j.bios.2014.04.042

[80] Smietana M, Bock WJ, Mikulic P, Ng A, Chinnappan R, Zourob M. Detection of bacteria using bacteriophages as recognition elements immobilized on long-period fiber gratings. Optics Express. 2011;**19**(9):7971-7978. DOI: 10.1364/OE.19.007971

[81] Koba M, Smietana M, Brzozowska E, Gorska S, Mikulic P, Bock WJ. Reusable bacteriophage adhesin-coated long-period grating sensor for bacterial lipopolysaccharide recognition. Journal of Lightwave Technology. 2015;**33**(12):2518-2523. DOI: 10.1109/JLT.2014.2364118

[82] Chen X, Zhang L, Zhou K, Davies E, Sugden K, Bennion I, et al. Real-time detection of DNA interactions with long-period fiber-grating-based biosensor. Optics Letters. 2007;**32**(17):2541-2543. DOI: 10.1364/OL.32.002541

[83] Chen X, Liu C, Hughes MD, Nagel DA, Hine AV, Zhang L. EDC-mediated oligonucleotide immobilization on a long period grating optical biosensor. Biosensors and Bioelectronics. 2015;**6**(2):1-6. DOI: 10.4172/2155-6210.1000173

[84] Hromadka J, Tokay B, James S, Tatam RP, Korposh S. Optical fibre long period grating gas sensor modified with metal organic framework thin films. Sensors and Actuators B: Chemical. 2015;**221**:891-899. DOI: 10.1016/j.snb.2015.07.027

[85] Smietana M, Bock WJ, Chen J, Mikulic P. Highly sensitive pressure

sensor based on long-period gratings written in boron co-doped optical fiber. Measurement Science and Technology. 2010;**21**(9):1-5. DOI: 10.1088/0957-0233/21/9/094026

[86] James SW, Korposh S, Lee SW, Tatam RP. A long period grating-based chemical sensor insensitive to the influence of interfering parameters. Optics Express. 2014;**22**(7):8012-8023. DOI: 10.1364/OE.22.008012

[87] Garg R, Tripathi SM, Thyagarajan K, Bock WJ. Long period fiber grating based temperature-compensated high performance sensor for bio-chemical sensing applications. Sensors and Actuators B: Chemical. 2013;**176**: 1121-1127. DOI: 10.1016/j.snb.2012.08.059

Chapter 8

Whispering Gallery Modes for Accurate Characterization of Optical Fibers' Parameters

Martina Delgado-Pinar, Xavier Roselló-Mechó, Emmanuel Rivera-Pérez, Antonio Díez, José Luis Cruz and Miguel V. Andrés

Abstract

Whispering gallery modes (WGMs) are surface modes that propagate azimuthally around resonators with rotational symmetry (toroidal, spherical, or, as in our case, cylindrical shaped, since the optical fiber itself plays the role of the microresonator). These modes are resonant in optical wavelength, and the spectral position of the resonances depends on the radius and the refractive index of the microresonator material. Due to the high-quality factor of the resonances (as high as 10^7 in cylindrical microresonators), they allow measuring different parameters with high sensitivities and very low detection limits. Here, we report the use of WGMs to characterize the properties of the material that forms the microresonator. In particular, we highlight the use of this technique to measure temperature profiles along conventional and special fibers (such as photosensitive or doped fibers), elasto-optic coefficients, and UV-induced absorption loss coefficients of different photosensitive fibers. These parameters of the fibers set the optical response of fiber-based components and may change when the device is in use in an optical system; thus, this technique allows an accurate characterization of the devices and leads to proper designs of components with specific optical responses.

Keywords: whispering gallery modes, surface modes, microresonators, optical fibers, fiber Bragg gratings, elasto-optic effect, thermo-optic effect

1. Introduction

Whispering gallery modes are surface modes that propagate azimuthally around resonators with rotational symmetry, generally a dielectric. This phenomenon was first described by Lord Rayleigh in the nineteenth century, when studying the propagation of acoustic waves in interfaces with a curvature [1]. St. Paul's Cathedral (London, UK), the Temple of Heaven (Beijing, China), the Pantheon (Rome, Italy), the Tomb of Agamemnon (Mycenae, Greece), and the Whispering Gallery in the Alhambra (Granada, Spain) are examples of architectonical structures that support acoustic modes which propagate guided by the surface of the walls. It was at the beginning of the twentieth century when the study of this guiding mechanism was extended to the electromagnetic waves, since Mie developed his theory for the plane electromagnetic waves dispersed by spheres with diameters of the same size as the optical wavelength [2]. Shortly after, Debye stablished the equations for the optical resonances of dielectric and metallic spheres based on Mie's dispersion theory [3]. The detailed study of the mathematical equations of WGMs was performed by Richtmyer [4] and Stratton [5], who predicted high-quality factors Q for these resonances and led to its implementation in different technologies based on microwave and acoustic waves. In the microscopic world, light can be guided by the same mechanism, when the resonator has dimensions of tens to hundreds of microns, and the wavelength of the light is in the visible-infrared range. In 1989, Braginsky et al. set the beginning of the optical WGMs when reporting the technique to excite optical modes in microresonators with spherical shape [6]. Since then, many researchers have studied the propagation of WGMs in structures with different symmetries [7] and have reported efficient methods based on microtapers to excite these modes in the optical range [8].

Due to the intrinsic low losses, WGMs show very high Q factors. For example, they can achieve values of 10^{10} in spheres [9], 10^{8} in silicon microtoroids [10], or 10^{6}–10^{7} in cylindrical microresonators [11]. At the resonance, the light guided by a WGM is recirculated in the microresonator many times, which provides a mechanism for decreasing the detection limit of the sensors based on them. This enhanced detection limit has been demonstrated to be low enough to measure a single molecule on the surface of a microtoroid [12].

WGM resonances shift in wavelength as the refractive index of the external medium changes. The sensitivity of WGMs as a function of these variations is significant: when considering a silica-cylindrical microresonator of 125 μm in diameter, immersed in water ($n = 1.33$), the calculated shift in wavelength of the resonance is 77 nm/RIU. For a typical resonance width of 0.5 pm, this leads to a detection limit of 6×10^{-6} RIU. It is worth to note that the light guided by WGMs is mainly confined in the microresonator. Thus, their sensitivity to variations of the material refractive index will be even higher. For example, it can achieve values as high as 1.1 μm/RIU when considering variations of the refractive index of the silica. In this example, the detection limit of the WGM decreases down to 4×10^{-7} RIU. In this chapter, we will report the use of WGMs in silica, cylindrical microresonators (an optical fiber) to measure and characterize the properties of the microresonator itself. There are a number of parameters, such as temperature or strain, which modify the refractive index of the material. Thus, this technique allows measuring with accuracy variations of temperature in doped optical fibers, in optical devices as fiber Bragg gratings (FBG), the elasto-optic coefficients of conventional silica fibers, and the absorption coefficient of photosensitive optical fibers, for example. We will report here the fundamentals of the technique, as well as the experimental results we obtained for these experiments.

2. Fundamentals

The guiding mechanism of WGMs in the azimuthal direction of a microresonator (MR) is total internal reflection, just as in the case of axial propagation in a conventional waveguide; see **Figure 1a**. Resonance occurs when the guided wave travels along the perimeter of the MR, and it drives itself coherently by returning in phase after every revolution. In its way, the wave follows continuously the surface of the MR, and the optical path in a circumnavigation must be equal to an integer multiple of the optical wavelength, λ. When this condition is fulfilled, resonances appear, and a series of discrete modes at specific wavelengths will show up. The resonant condition can be written as [13]

$$\lambda_R = \frac{2\pi a \cdot n_{\text{eff}}}{m} \tag{1}$$

where λ_R is the resonant wavelength, a is the radius of the MR, n_{eff} is the effective index of the WGM, and m is the azimuthal order of the mode (i.e., the number of wavelengths in the perimeter of the MR). The effective indices of the different modes are calculated, as usual, by solving Maxwell's equations and applying the proper boundary conditions [5]. In our case, we will deal with cylindrical, dielectric MRs with translational symmetry in the axial direction (see **Figure 1b**). Two zones can be identified, regions I (of radius a) and II (which extends to the infinite), with refractive indices n_1 and n_2, respectively, with $n_1 > n_2$. The magnetic permeability of the material and of the external medium is equal to that of the vacuum, μ_0, and both media are homogeneous although, in general, they present an anisotropy in the dielectric permittivity. In the axial direction, we will consider a refractive index of the material n_{1z} which is different to the refractive index in the transversal directions, n_{1t} (see Eq. (2) for the expression of the tensor of the refractive index):

$$n_2 = \begin{pmatrix} n_{1t} & 0 & 0 \\ 0 & n_{1t} & 0 \\ 0 & 0 & n_{1z} \end{pmatrix} \tag{2}$$

We do not intend to give a full description of the solution of this problem, which can be found in [14], but we will summarize the main equations and features of WGMs.

If we solve Maxwell's equations with this uniaxial tensor, the modes split in two series of family modes that, analogously to the case of axial waveguides, are denoted as TE-WGMs, which show a transversal electric field ($e_z = 0$), and TM-WGMs, with transversal magnetic field ($h_z = 0$). Each series of modes is ruled by a transcendental equation that must be solved: Eq. (3) for TM modes and Eq. (4) for TE modes. The solutions consist on a series of discrete wavelengths, which correspond to the different radial orders l of each mth value. With these values, it is possible to calculate the effective indices of each WGM resonance using Eq. (1):

$$n_{1z}\frac{J_{m'}(k_0 n_{1z} a)}{J_m(k_0 n_{1z} a)} = n_2 \frac{H_m^{(2)'}(k_0 n_2 a)}{H_m^{(2)}(k_0 n_2 a)} \tag{3}$$

$$\frac{1}{n_{1t}}\frac{J_{m'}(k_0 n_{1t} a)}{J_m(k_0 n_{1t} a)} = \frac{1}{n_2}\frac{H_m^{(2)'}(k_0 n_2 a)}{H_m^{(2)}(k_0 n_2 a)} \tag{4}$$

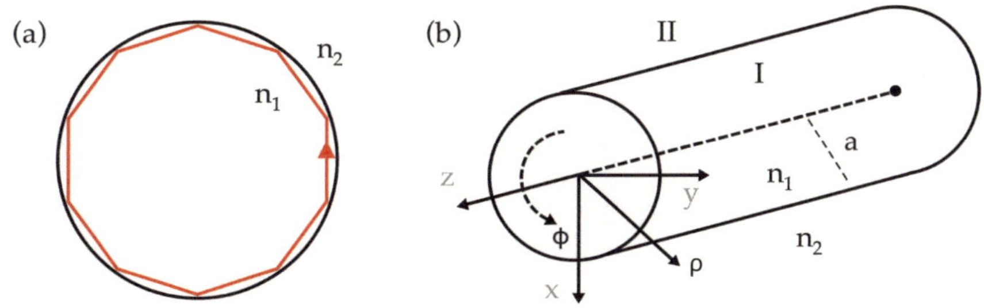

Figure 1.
(a) Scheme of the WGM propagating azimuthally in the MR. (b) Cylindrical system of coordinates which shows the two regions considered in the problem.

In Eqs. (3) and (4), k_0 is the wavenumber in vacuum, $k_0 = 2\pi/\lambda$, J_m is the Bessel function of order m and J'_m is its first derivative, $H_m^{(2)}$ is the second class Hankel function of order m, and $H_m^{(2)'}$ is its first derivative. We have considered that the external medium does not present any anisotropy (in our case, it will be air).

By following this procedure, it is possible to calculate the dispersion curves of several WGMs propagating in a cylindrical, silica MR of 125 μm diameter (the parameters of conventional optical fibers). Sellmeier dispersion of the silica was taken into account for the refractive index of the material. It is worth to note that the dispersion curves are not truly a curve, but a series of discrete solutions that have a particular radial order l and azimuthal order m. For a standard optical fiber and 1550 nm optical wavelength, the azimuthal orders will be relatively high ($m \sim 300$). **Figure 2** shows the calculations of the resonant wavelengths for the first radial orders, as a function of the azimuthal order m for the TM polarization. The curves for the TE polarization follow the same trend, but the values of the resonant wavelengths are slightly different. By using Eq. (1), it is possible to relate the resonant wavelength with the effective index of the WGM resonance. For the azimuthal order $m = 360$ and the first radial order, $l = 1$, the resonant wavelengths and the effective indices of both polarization families are $\lambda_R^{TM} = 1508.25$ nm, $n_{eff}^{TM} = 1.3826$ and $\lambda_R^{TE} = 1505.39$ nm, $n_{eff}^{TE} = 1.3800$. Thus, the resonances for each polarization are not overlapped in wavelength.

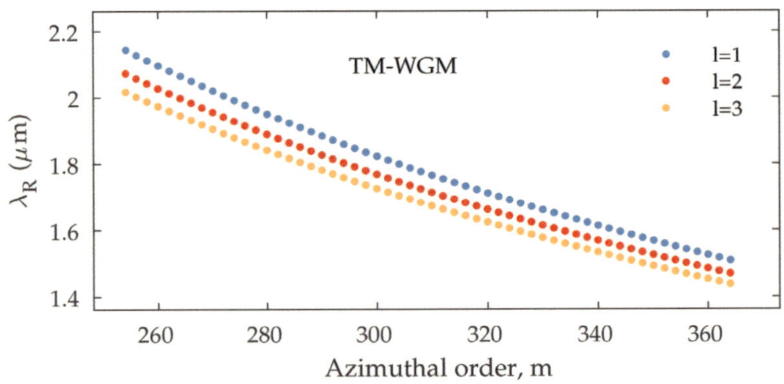

Figure 2.
Resonant wavelength of WGMs with azimuthal orders from 250 to 370. Only a selection of the solutions to highlight their discrete nature is shown.

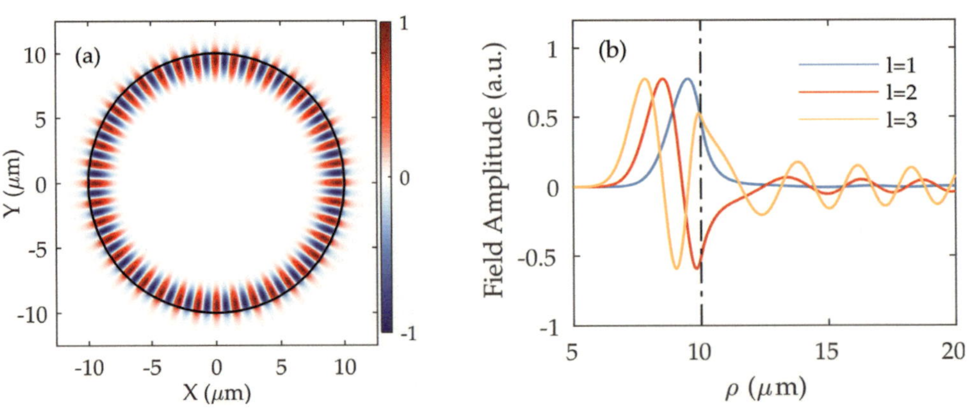

Figure 3.
(a) Optical field of a $m = 40$ and $l = 1$ WGM in a silica, cylindrical MR. (b) Field amplitude of the WGM as a function of the radial coordinate, for $m = 40$ and $l = 1, 2, 3$.

Regarding the distribution of the fields, **Figure 3a** shows the amplitude of the electric field of the first radial order TM-WGM, propagating in a cylindrical, silica MR of 10 μm diameter (the order m of the mode is 40; a low-order mode was considered in order to show the details of the field). As it can be observed, the field is well confined within the MR material (although its evanescent field is high enough to enable the use of these modes for sensing). As the azimuthal order of the WGM gets higher, the field will be more localized near the interface between the MR and the external medium. Also, it should be noted that, as the radial order of the WGM increases, the evanescent tail in the outer medium is larger; thus the quality factor of the correspondent resonance will be poorer. **Figure 3b** shows the field amplitude along the radial coordinate of the MR. As it can be observed, the optical power is localized in the outer region of the MR, near the interface, and shows a low-evanescent field in the outer medium, especially for the $l = 1$ mode.

3. Experimental setup

The general setup used in the experiments is shown in **Figure 4a**. The light source is a tunable diode, linearly polarized laser (TDL) with a narrow linewidth (<300 kHz). The tuning range covers from 1515 to 1545 nm. The laser integrated a piezoelectric-based fine frequency tuning facility that allows continuous scanning of the emitted signal around a given wavelength, with subpicometer resolution. A polarization controller (PC) after the laser allows rotating the polarization of the light, and, as a consequence, it allows exciting TE- and TM-WGMs separately. The optical signal is then launched through an optical circulator, which enables measuring the WGM resonances in reflection by means of a photodetector (PD).

The MR will consist on a section of the bare optical fiber under test (FUT). Depending on the experiment, it will be a conventional telecom fiber, a rare-earth doped fiber, a photosensitive fiber, or a fiber where a grating has been previously inscribed. It is carefully cleaned and mounted on a three-axis flexure stage. WGMs are excited around the FUT by using the evanescent optical field of an auxiliary microtaper with a waist of 1–2 μm in diameter and a few millimeters in length. This is not the only method that allows exciting WGMs in MRs: for example, one of the first techniques consisted on using a prism to excite the resonances in a spherical MR [15], but the efficiency was very poor. More recently, a fused-tapered fiber tip fabricated using a conventional fiber splicer was demonstrated to be capable of exciting WGMs in a cylindrical MR [16]. However, the highest efficiencies are achieved by using microtapers, with coupling efficiencies higher than 99% [8]. These microtapers are fabricated by the fuse-and-pull technique from conventional

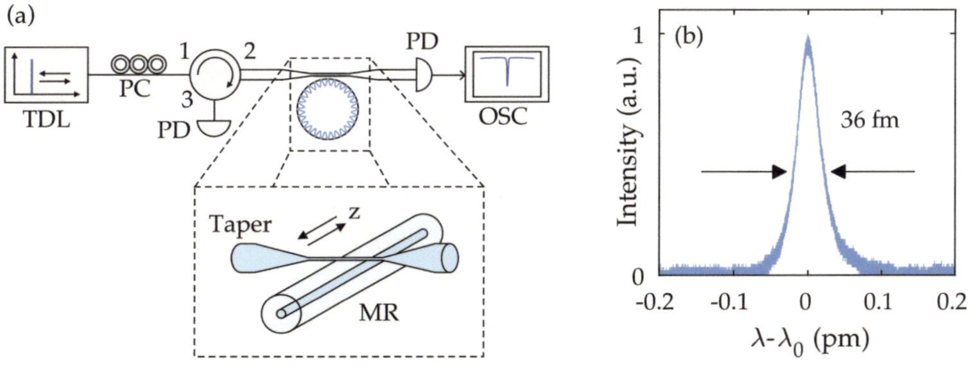

Figure 4.
(a) Scheme of the experimental setup. (b) Typical reflection spectrum of a WGM.

telecom fiber [17]. The microtaper and the MR are placed perpendicularly (see in the inset **Figure 4a**). Since the optical field of the WGMs is not axially localized (its extension is around 200 μm in length [7]), this setup allows exciting the WGM at different positions along the MR: by sweeping the microtaper along the MR, it is possible to detect variations of the parameters of the MR in the axial direction by measuring the shift of the resonances—radius [13, 18], temperature, or strain. Variations can be characterized along several centimeters of the MR.

The transmission of the taper was measured using a photodetector, and the signal was registered by an oscilloscope synchronized with the TDL. A typical transmission trace consists on a signal that will present a series of notches at the resonant wavelengths. For MRs of 125 μm in diameter, the free spectral range between two consecutive azimuthal orders m is $\sim$ 4 nm at 1550 nm, and it is the same for both polarizations. **Figure 4b** shows the reflection spectrum of a resonance in an optical fiber ($a = 62.5\ \mu m$): its linewidth is 36 fm, which corresponds to a Q-loaded factor of 4×10^7.

As it was mentioned before, the position of the resonances will depend on the value of the refractive index of the material. In the next sections, we will study the characterization of different fibers and fiber components by means of the measurement of the shift of WGM resonances as the effective index of the MR is modified.

4. Measurement of temperature profiles in doped fibers and fiber gratings

When a silica fiber is heated up, two effects occur. First, the expansion of the fiber leads to a change of the diameter. Second, the thermo-optic effect induces a change in the refractive index of the material due to a variation of temperature. This variation modifies the spectral position of the WGM. From Eq. (1) it is possible to evaluate the shift of the resonant wavelength, $\Delta\lambda_R$, of a WGM due to a variation of temperature, ΔT:

$$\frac{\Delta\lambda_R}{\lambda_R} = \left(\frac{1}{a}\frac{da}{dT} + \frac{1}{n_{\text{eff}}}\frac{dn_{\text{eff}}}{dT}\right) \cdot \Delta T \tag{5}$$

In the case of optical fibers as MRs, it is a good approximation to assume that the thermo-optic coefficient (i.e., the second term in Eq. (5)) can be replaced by that of the pure silica, since the optical field of the WGMs is mainly localized in the fiber cladding (see **Figure 3**). The high sensitivity of WGMs to variations of temperature has been demonstrated for different geometries of the MR, such as microspheres [19, 20] or cylinders [21]. Moreover, the propagation of an optical signal of moderate power ($\sim$ 1 W or higher) in a fiber generally induces a variation of temperature of the material. Due to the variation of temperature, the optical response of the fibers, or fiber components, may change when they are in operation. Thus, a detailed characterization of this effect is of interest to design properly the fiber-based optical systems. The use of WGMs allows achieving a very low detection limit: Rivera et al. claimed a detection limit of two thousandths of degree [21].

Here, we will present the characterization of temperature variations in two different examples: (i) rare-earth doped active fibers and (ii) fiber gratings inscribed in commercial photosensitive fibers.

4.1 Measurement of temperature in rare-earth doped fibers

Heating of rare-earth doped fibers can be an issue in fiber-based lasers and amplifiers. For example, thermal effects can be a limit to the maximum output power that these systems can provide [22]. Another example is the shift in wavelength observed in distributed Bragg reflectors (DBR) and distributed feedback (DFB) lasers due to a pump-induced increment of temperature [23]. The heat is due to the non-radiative processes related to the electronic relaxation of some dopants: for example, this effect is less important in ytterbium-doped fibers, while Er/Yb-codoped and erbium-doped fibers exhibited a high increase of temperature with pump, due to its specific electronic-level system [24]. Thus, it is an intrinsic characteristic of the doped fibers that one needs to evaluate in order to design the proper optical system.

In the experiments presented here, several commercially available single-mode, core-pumped doped fibers from Fibercore were investigated. Specifically, the FUTs were three Er-doped fibers (DF-1500-F-980, M12-980/125, and I25-980/125), a Yb-doped fiber (DF-1100), and an Er/Yb-codoped fiber (DF-1500 Y). The values for absorption coefficients at the pump wavelength were 5.5 dB/m (DF-1500-F-980), 12 dB/m (M12-980/125), 21.9 dB/m (I25-980/125), 1000 dB/m (DF-1100), and 1700 dB/m (DF-1100). Short sections of $\sim$ 2 cm in length of each FUT were used as the MR where the WGMs were excited. The FUTs were pumped with a single-mode, fiber-pigtailed laser diode that emitted a maximum power of 380 mW at 976 nm. As the pump launched to the FUT was increased, the WGM resonance shifted toward longer wavelengths in all cases, as it was expected, since the thermo-optic and the thermal expansion coefficients of silica are both positives. As an example, **Figure 5** shows the shift in wavelength of a resonance as a function of the pump launched to the fiber DF-1500-F-980. In our experiments, we did not investigate in detail the temporal response of the phenomenon, which will be ruled by the mechanisms that convert the pump power to heat, the heat conduction in silica, and the transfer of heat to the air. Typically, it will be on the range of a few tens of microseconds [25].

At this point, several features of this technique must be clarified. First, it is worth to point out that the shift in wavelength is virtually independent of the particular resonance used for the measurements, that is, it does not depend on its radial and azimuthal order nor on its polarization. The sensitivity to thermal variations of different WGM resonances was theoretically calculated around 1.53 μm, taking into account both the thermal expansion of the fiber and the thermo-optic effect. The results showed that the difference in sensitivity between different resonances differs in less than 1/10000 per each ^oC of temperature increase. This simplifies the utility of this technique.

The second aspect to highlight is related to the fact that the dopants in the active fibers are located in their core, while WGMs are highly confined in the outer region of the cladding (see **Figure 3**). From the study of heat conduction in doped fibers

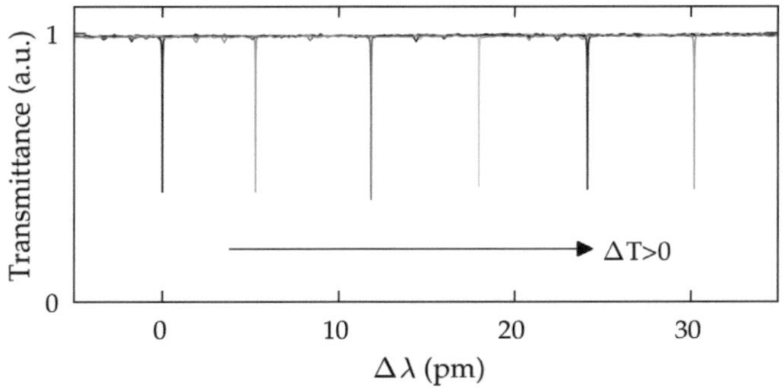

Figure 5.
Wavelength shift of the resonant wavelength as the pump power is increased. From left to right: pump power 0 mW, 40 mW, 110 mW, 180 mW, 270 mW, and 370 mW.

carried out by Davis et al. [25], it is possible to calculate that, at the steady state, the increase of temperature at the core of the fiber is just 1.5% larger than at the outer surface.

In order to calibrate the shift in wavelength of the WGM resonances with the heating, a FBG inscribed in the core of a doped fiber was used for comparison. The procedure is described in [21]. The WGM resonances shift at a rate of 8.2 pm/^oC. With this calibration, it is possible to correlate the shifts in wavelength with the increase of temperature in the core of the fiber. For the example shown in **Figure 5**, the maximum increment of temperature achieved for a pump of 370 mW was 3.7^oC.

Figure 6 summarizes the measurements performed for the different doped fibers. A similar trend can be observed in all the cases; the resonances shift fast in wavelength for low pump powers, and, beyond certain pump, heating tends to saturate. It can be observed that the Yb fiber DF 1100 shows a similar increase of temperature to those of the Er-doped fibers, although the concentration of the dopants in the Yb fiber is much larger (note the absorption coefficient around 975 nm). Also, the highest temperature increment corresponds to the Er/Yb-doped fiber (DF 1500 Y), despite that it shows a lower absorption coefficient than its equivalent Yb-doped fiber (DF 1100). These results are in accordance to the fact that the heating is related to the existence of non-radiative transitions for the relaxation of electrons in the active medium.

4.2 Measurement of temperature profiles in fiber components

As it was mentioned before, WGMs are axially localized: their extension along the fiber is $\sim 200\ \mu m$, typically, for a MR of 62.5 μm. Thus, this technique provides spatial resolution. The taper can be swept along the MR in order to characterize the parameters of the FUT point to point. This feature was used in order to characterize the temperature profile along fiber components [26].

The FBGs used in the experiments were written in germanium-silicate boron codoped, photosensitive fibers from Fibercore, using a doubled-argon UV laser and a uniform phase mask. The length of all the gratings was $\sim$ 10 mm. The WGMs were excited at different positions along the FBG, and, simultaneously, it was illuminated by optical signals of moderate powers, within or outside of the

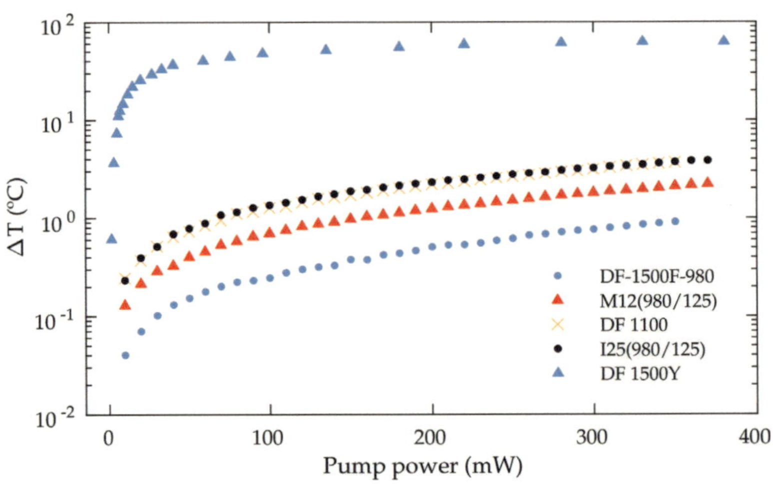

Figure 6.
Heating of the doped fibers as a function of pump power.

reflection band but in the vicinities of the Bragg wavelength. This illumination signal was provided by an amplified tunable laser (range, 1520–1560 nm) that provided up to 1 W of CW light.

As a preliminary experiment, a section of fiber Fibercore PS980 was uniformly irradiated (i.e., there was no grating inscribed). The length was 5 mm, and the UV fluence power used in the irradiation was 150 J/mm^2. The wavelength shift of the resonances was measured as the MR was illuminated with a 1550 nm optical signal, compared to the original position of the resonances, with no illumination along the FUT. **Figure 7a** shows the results. The data show a clear difference between the irradiated length ($z < 3$ mm) and the non-irradiated length ($z>3$ mm). A temperature gradient in an intermediate region due to the heat conduction in silica and the transfer of heat to the air can be observed. It should be noted that this section is far larger than the length of the focused UV beam ($\sim 700\ \mu m$); thus, the beam size is not the cause of this transition length. In the irradiated section, the temperature increases at a rate higher than $10\ {}^oC/W$, for this sample, while the pristine fiber heats up at a rate lower than $1\ {}^oC/W$. The increment of temperature was linear with power in the available power range. This experiment avails that this technique allows characterizing the variations of temperature along the components with a resolution of tenths of a millimeter. This feature is useful when one needs to detect, evaluate, and correct smooth undesired non-homogeneities that may occur during the fabrication of FBG and LPG, which are usually short components. As an example, **Figure 7b** shows the measurement of the temperature profile of a section of an irradiated fiber (length, 5 mm) that suffered from some misalignment during the UV irradiation process. For this sample, a variation of $4\ {}^oC$ is measured in such a short irradiated length.

The temperature profile along a FBG with strong reflectivity was measured using this technique. The FBG had a reflectivity higher than 99.9%; the Bragg wavelength was 1556 nm, its length was 12 mm, and it was fabricated in PS1250 fiber (Fibercore). First, the illumination signal was tuned well outside the reflection band, at 1540 nm; in this case, there is no reflection of the optical signal; it just propagates through the FBG. The power launched to the MR was 800 mW. Curve (i) in **Figure 8** shows the obtained results. As expected, a similar result to the case shown in **Figure 7a** was obtained: the heating over the length of the FBG was fairly constant, $\sim 5.5\ {}^oC$. It should be noted that the axial resolution of the technique will be larger than the grating period. Then, the average increment of temperature should be similar to that introduced in the case of the uniformly irradiated fiber, for

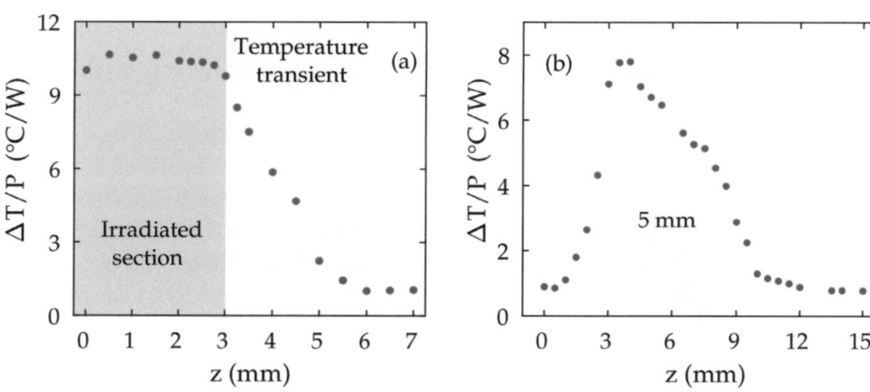

Figure 7.
Temperature profile of irradiated FUTs, (a) uniformly and (b) nonuniformly.

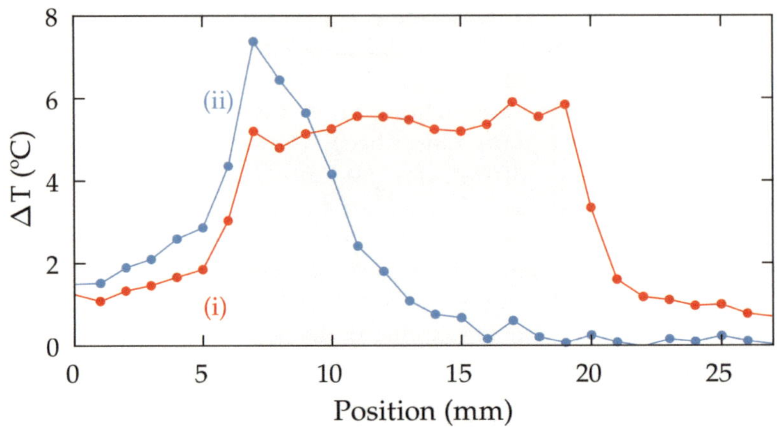

Figure 8.
Temperature profile of a FBG illuminated (i) outside and (ii) within the reflexion band.

the same UV fluence and fiber characteristics. Two transition zones were clearly observed at both ends of the grating.

Finally, the temperature profile was measured when the optical signal was tuned to the Bragg wavelength (power, 1 W) (see curve (ii) in **Figure 8**). In this case, one should take into account that the UV irradiation is constant over its length, and the gradient temperature is due to the fact that the optical signal is reflected as it penetrates into the grating. A sharp increment of temperature at the beginning of the grating, at the extreme that is illuminated, can be observed. The maximum is located at the vicinities of the point where the FBG begins. The decay of temperature extends over a length of ~ 5 mm, which is shorter than the length of the fiber itself (12 mm). This is consistent with the high reflectivity of this FBG. Moreover it should be noted that, at the beginning of the curve, that is, $z = 0 - 3$ mm, the temperature increase is $\sim 2\ ^oC$, that is, roughly twice the value obtained for a pristine fiber. On the contrary, in the section after the grating (and even at the last millimeters of the FBG), the increment of temperature is below the detection limit of the technique. The origin of this asymmetry is the reflection of the optical signal: the amount of light that reaches the last millimeters of the FBG is very small. This technique, then, provides information about the effective length of gratings of different reflectivity, information that could be relevant for the design of optical systems that require of short cavities, or cavities that require of a very precise length, as in the case of mode-locked fiber lasers.

5. Measurement of absorption coefficients in photosensitive fibers

In the previous section, the gradient of temperature induced in fiber-optic components by means of an illumination signal has been characterized and discussed. It has been shown that there is a difference in temperature between the sections that have been irradiated with UV light compared to the pristine fibers. It is well known that the UV irradiation induces a change in the index of photosensitive fibers, which is employed to fabricate FBGs and LPGs. According to Kramers-Kronig relations, the change in the refractive index is associated with a variation of the absorption coefficient. In addition, the exposure of the fiber to the levels of UV light usually employed in the grating fabrication induces mechanical deformations in the fiber [27]. This leads to an increase of the loss due to scattering. Thus, when a fiber is UV irradiated, its loss, α, increases due to two causes: absorption that will be quantified by α^{abs} and scattering, α^{scat}.

The increase of α introduced by UV irradiation has been measured before [28], since this is a parameter of interest to optimize the fabrication of FBGs, especially in the case of long or superimposed gratings with many reflection bands [29, 30]. This measurement provides information about an averaged value of the attenuation loss along the irradiated section, which includes both the absorption and the scattering contributions. The technique based on the measurement of the shift of WGM resonances will only measure the absorption coefficient; thus, by combining the two types of measurements, it is possible to evaluate both contributions separately.

Different types of photosensitive fibers were studied [11]: (i) Fibercore PS980, (ii) Fibercore PS1250, (iii) Fibercore SM1500, and (iv) Corning SMF28; this fiber was hydrogenated for 15 days (pressure: 30 bar) to increase its photosensitivity. The setup used in the experiments was the same than in the previous experiments shown in this chapter. In this case, the FUTs were short sections of the different fibers, which were exposed to a UV fluence of 150 J/mm^2. Similar temperature profiles to that shown in **Figure 7a** were obtained for all of them, but with different temperature increments, since the photosensitivity was also different for each of them.

The different increases of temperature between the irradiated fiber and the pristine fiber will provide us information to quantify the variation in the α^{abs} due to UV irradiation. It will be assumed that the heating over the transversal section of the fiber, at a given axial position, is set by the absorption coefficient, α^{abs}. According to the analysis reported by Davis et al. [25], the heating at the steady state, ΔT, will be given by

$$\frac{\Delta T}{P} = \frac{1}{2\pi a h} \alpha^{abs} \tag{6}$$

where h is the heat transfer coefficient ($81.4\ W \cdot m^{-2} \cdot K^{-1}$ for a silica fiber). Then, the ratio of ΔT between two different points along the FUT, 1 and 2, is given by

$$\frac{\Delta T_2}{\Delta T_1} = \frac{\alpha_2^{abs}}{\alpha_1^{abs}} \tag{7}$$

Thus, with this analysis and the experimental data obtained from the measurement of the wavelength shift of WGM resonances in irradiated points (1) and pristine points (2) of the FUT, this ratio between the respective α^{abs} can be calculated.

Direct measurements of transmission loss variation as the fibers were irradiated were carried out for a PS980 fiber. First, the value of the loss of the pristine fiber was measured at 1550 nm by means of the cutback method: the obtained value was 120.0 ± 0.5 dB/km. Then, the UV laser was swept back and forth along a 5-cm-long section of the fiber, repeatedly. The full description of the procedure is described in [11]. **Figure 9a** shows the data obtained in this experiment. The final loss was 6.2 ± 0.4 dB/m; thus the ratio between the loss coefficients, α_2/α_1, increased 52 ± 3 times. Please remember that this loss coefficient includes both absorption and scattering contributions ($\alpha = \alpha^{abs} + \alpha^{scat}$).

The contribution to the loss by means of the absorption mechanism was measured using the WGM technique (see **Figure 9b**). In this case, a 1550 nm laser (maximum power, 1 W) was launched to the FUT, and the thermal shift of the

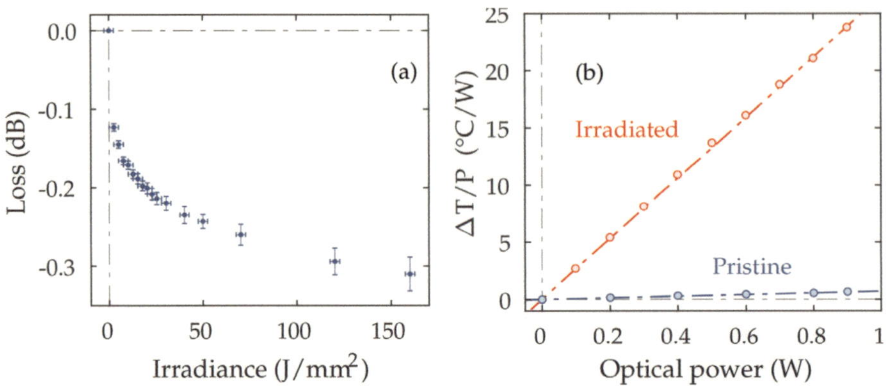

Figure 9.
(a) Direct measurement of the loss as the PS980 fiber is irradiated. (b) Heating of the PS980 fiber as a function of the illumination power.

	WGM technique			Direct measurements		
	$1-\alpha_2^{abs}/\alpha_1^{abs}$	$\Delta T/P$ (°C/W)		α_2/α_1	α (dB/km)	
		Irradiated	Pristine		Irradiated	Pristine
PS980	36.9 ± 0.7	26.48 ± 0.15	0.718 ± 0.014	52 ± 3	6200 ± 400	120 ± 2[3]
PS1250	40.1 ± 0.8	30.80 ± 0.17	0.768 ± 0.014	50 ± 3	6600 ± 400	131.1[4]
SM1500	>40[1]	1.20 ± 0.03	< 0.03[2]	190 ± 50	370 ± 90	1.95[4]
H_2-SMF28	28.8 ± 0.5	23.48 ± 0.13	0.815 ± 0.012	n/a[2]	5600 ± 400	n/a[1]

[1]*Nonavailable.*
[2]*Below detection limit.*
[3]*Cutback measurement.*
[4]*Nominal value.*

Table 1.
Measurement of thermal heating and loss coefficient of different fibers.

resonances was measured as the laser power was increased, at two different points, one within the irradiated section and one outside it. The data does not show any sign of saturation of the heating, at this range of power. The temperature of the irradiated section increased linearly, at a rate of 26.48 ± 0.15 °C/W, and at 0.718 ± 0.014 °C/W in the pristine region. The ratio between these values, that is, the ratio $\alpha_2^{abs}/\alpha_1^{abs}$, is 36.9 ± 0.7 °C/W.

This process was repeated for all the different fibers mentioned before: PS1250, SM1500, and hydrogenated SMF28, at 1550. **Table 1** includes the results from the measurements and the corresponding analysis: α_2/α_1 was obtained for each of them from the direct measurement of the loss, while $\alpha_2^{abs}/\alpha_1^{abs}$ was calculated from the technique based in WGMs.

The results, compiled in **Table 1**, allow establishing several conclusions of interest. First, as expected, the absorption coefficient is substantially increased due to the UV irradiation. As a consequence, even for signals of moderate powers, FBGs might experience shifts and chirps that should be taken into account [31]. Second, the results show that α_2/α_1 is systematically higher than $\alpha_2^{abs}/\alpha_1^{abs}$. Roselló-Mechó et al. analyzed the measurements to demonstrate that these results lead to the conclusion that scattering loss increases at a higher rate than absorption loss [11].

Finally, Eq. (6) can be used to calculate the absolute value of the absorption and scattering coefficients by taking into account the values of h and a for a silica

	α^{abs} (dB/km)		α^{scat} (dB/km)	
	Irradiated	**Pristine**	**Irradiated**	**Pristine**
PS980	3680 ± 20	99.7 ± 1.9	2500 ± 400	20 ± 3
PS1250	4280 ± 20	106.6 ± 1.9	2300 ± 400	24.5 ± 1.9
SM1500	167 ± 4	< 1.95[1]	200 ± 90	< 1.95[1]
H_2-SMF28	3260 ± 18	113.1 ± 1.7	2300 ± 400	n/a[2]

[1]*Nominal value.*
[2]*Nonavailable, hydrogenated fiber.*

Table 2.
Absorption and scattering contributions to the overall attenuation coefficient.

fiber [25] and the measurements of α. The results of the contributions are compiled in **Table 2**. Both contributions are in the same order of magnitude, but α^{scat} is smaller for three of the four FUTs. These values confirm that scattering loss increases faster than absorption loss.

Thus, by means of the combination of both techniques, it is possible to quantify the different contributions to the loss, even for short sections of fiber. This information might be useful, for example, in the design of novel-active doped fibers, since it is possible to evaluate if the doping technique increases the scattering loss unnecessarily, but not so much the absorption.

6. Measurement of Pockels coefficients in optical fibers

The elasto-optic effect consists on the variation in the refractive index generated by any strain applied to the fiber. The correspondent elasto-optic coefficients are usually determined by measuring the optical activity induced by a mechanical twist and the phase change induced by longitudinal strain [32, 33]. This technique relies on the use of the conventional axial modes propagating through the fiber. Since these modes are essentially transverse to the axis of the fiber [34], the anisotropy of the elasto-optic effect does not show up. On the contrary, WGMs have a significant longitudinal component; hence, their optical fields experience the anisotropy of the elasto-optic effect intrinsically. In the last years, researchers have demonstrated a number of fiber devices in which the longitudinal components of the electromagnetic modes are significant, such as microfibers [35] and microstructured optical fibers with a high air-filling fraction [36]. For these cases, the measurement and characterization of the anisotropy of the elasto-optic effect and its Pockels coefficients are of high interest. Roselló-Mechó et al. reported a technique based on the different wavelength shifts of TE- and TM-WGM resonances in a fiber under axial strain, to measure these coefficients [37]. This technique has the additional advantage that, since it does not involve the conventional modes of the fiber, there is no need that the FUTs are single mode in order to carry out the measurements. Then, the coefficients can be measured at different wavelengths to determine their dispersion; this is a limitation of the usual technique based on the optical activity which is overcome by means of WGM technique [38].

According to Eq. (1), a variation in the refractive index will tune the WGM resonances in wavelength. In this case, an axial strain will be applied to the FUT in order to induce this variation in the index, due to the elasto-optic effect. This feature was applied in different works in order to tune the WGM resonances

[39, 40]. However, there was not any mention to the different behaviors of TE- and TM-WGM.

An axial strain introduces a refractive index perturbation in an isotropic, cylindrical MR, due to the elasto-optic effect, which will be different for the axial (Δn_z) and transversal directions (Δn_t):

$$\frac{\Delta n_t}{n_0} = -p_{et}\,\varepsilon; \qquad p_{et} \equiv \frac{n_0^2}{2}\left(p_{12} - \nu(p_{11} + p_{12})\right), \tag{8}$$

$$\frac{\Delta n_z}{n_0} = -p_{ez}\,\varepsilon; \qquad p_{ez} \equiv \frac{n_0^2}{2}\left(p_{11} - 2\nu p_{12}\right), \tag{9}$$

where n_0 is the unperturbed index of the MR, p_{ij} are the elasto-optic coefficients, ν is Poisson's ratio, and ε is the strain applied to the MR. The coefficients p_{et} and p_{ez} are the effective elasto-optic coefficients, which are defined for simplicity. According to the reported values for the elasto-optic coefficients for fused silica ($p_{11} = 0.121, p_{12} = 0.27$ [41], $\nu = 0.17$ [42]), the ratio $\Delta n_t/\Delta n_z \approx 6.97$; hence, it is expected that the strain introduces a significant differential anisotropy. With this in mind, Maxwell's equations will be solved considering the uniaxial tensor given by Eq. (2). The solutions, as mentioned before, split in two families of WGM, the TE and TM modes, whose resonant frequencies will be obtained by solving Eqs. (3) and (4).

The refractive index perturbation is not the only factor to take into account when evaluating the wavelength shift of WGM resonances due to strain: the radius a of the MR also varies with it according to Poisson's ratio, $\Delta a/a = -\nu\varepsilon$.

With all these ideas in mind, the relative shift of the WGM resonances, $\Delta\lambda_R/\lambda_R$, can be characterized as a function of the strain, for TE- and TM-WGMs. **Figure 10a** shows an example of the anisotropic behavior of TE- and TM-WGM. The strain applied to the MR was 330 $\mu\varepsilon$ for both polarizations, and the measured wavelength shift was different for each of them: 0.18 nm for TE-WGM and 0.11 nm for TM-WGM. $\Delta\lambda_R/\lambda_R$ was measured as a function of the strain in detail at 1531 nm, for both polarizations; the results are shown in **Figure 10b**. A linear trend in both cases can be observed: the slopes of the linear regressions that fit the experimental values are $s_{TE} = -0.369 \pm 0.006\ \mu\varepsilon^{-1}$ for the TE- and $s_{TM} = -0.201 \pm 0.004\ \mu\varepsilon^{-1}$ for the TM-WGM. The ratio $s_{TE}/s_{TM} = 1.84$ shows the anisotropy of the elasto-optic effect. From these values, it is possible to calculate the elasto-optic coefficients p_{ij} with its uncertainties (see [37] for a more detailed description of the procedure), by taking

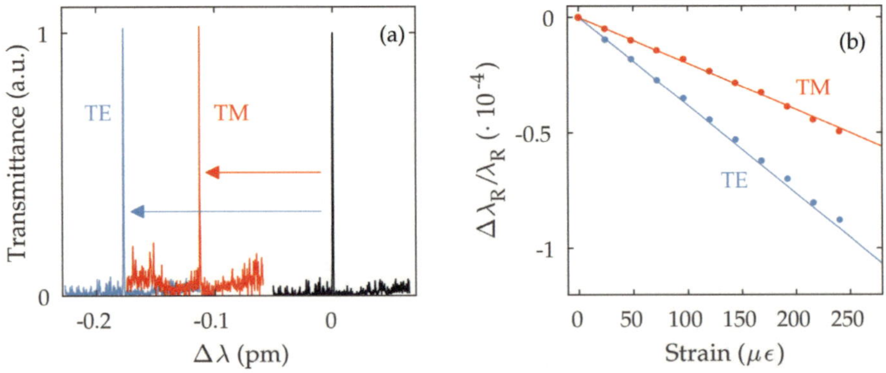

Figure 10.
(a) Wavelength shift of TE-/TM-WGM resonances for $\varepsilon = 330\ \mu\varepsilon$. (b) Measurement of the wavelength shift as a function of the strain.

	Present work		Literature
	1531 nm	1064 nm	
p_{11}	0.116	0.131	0.113 @ 633 nm [32]
			0.121 @ 633 nm [41]
p_{12}	0.255	0.267	0.252 @ 633 nm [32]
			0.270 @ 633 nm [41]

Table 3.
Comparison of experimental p_{ij} values with those reported in the literature.

into account the Sellmeier coefficients for the value of the refractive index at 1531 nm and Poisson's ratio of $\nu = 0.17 \pm 0.01$.

The measurements were repeated at 1064 nm, to study the dispersion of the elasto-optic effect. Results at both wavelengths are compiled in **Table 3** and are compared with those reported in the literature. Both sets of measurements are in good agreement, and the small differences might be due to the fact that the technique based in WGM measures the p_{ij} of the cladding material (i.e., fused silica), while in the case of the other techniques, the coefficients are determined by the material of the fiber core, which is usually silica doped with other elements.

7. Conclusions

In this chapter, we described a technique based on the excitation of WGMs around cylindrical MRs, to measure properties of the MR material. The resonant nature of the WGMs confers this technique with high sensitivity and low detection limits. Also, the technique allows measuring these parameters with axial resolution; hence, it is possible to detect changes of the parameters point to point along the MR.

The technique has been applied to different experiments. Mainly, thermo-optic effect and elasto-optic effect have been investigated in silica fibers. The variation in the index, due to a change in the temperature or strain, rules the shift in wavelength of the WGM resonances. When the technique was applied to different types of fibers and components, different information were obtained from the experiments. In particular, we measure temperature profiles in pumped, rare-earth doped fibers and in FBGs; the absorption coefficient in irradiated photosensitive fibers; and the Pockels coefficients in telecom fibers. Novel results were obtained: for example, it was possible to measure absorption and scattering loss coefficients separately, and, also, the anisotropy of the elasto-optic effect was observed experimentally. The information provided by the WGM-based technique might help to optimize the fabrication procedures of doped fibers and fiber components as FBGs or LPGs.

Acknowledgements

This work was funded by Ministerio de Economía y Competitividad of Spain and FEDER funds (Ref: TEC2016-76664-C2-1-R) and Generalitat Valenciana (Ref: PROMETEOII/2014/072), Universitat de València (UV-INV-AE16-485280). X. Roselló-Mechó's contract is funded by the FPI program (MinECo, Spain, BES-2014-068607). E. Rivera-Pérez's contract is funded by the Postdoctoral Stays in Foreigner Countries (291121, CONACYT, Mexico).

Author details

Martina Delgado-Pinar*, Xavier Roselló-Mechó, Emmanuel Rivera-Pérez, Antonio Díez, José Luis Cruz and Miguel V. Andrés

Applied Physics Department, Science Materials Institute, University of Valencia, Valencia, Spain

*Address all correspondence to: martina.delgado@uv.es

References

[1] Rayleigh L. The Theory of Sound. Vol. 2. London: Macmillan; 1896

[2] Mie G. Beiträge zur Optik trüber Medien, speziell kolloidaler Metallösungen. Annalen der Physik. 1908;**330**(3):377-445. DOI: 10.1002/andp.19083300302

[3] Debye P. Der Lichtdruck auf Kugeln von beliebigem Material. Annalen der Physik. 1909;**335**(11):57-136. DOI: 10.1002/andp.19093351103

[4] Richtmyer RD. Dielectric resonators. Journal of Applied Physics. 1939;**10**(6): 391-398. DOI: 10.1063/1.1707320

[5] Stratton JA. Electromagnetic Theory. 1st ed. New York: McGraw-Hill; 1941

[6] Braginsky VB, Gorodetsky ML, Ilchenko VS. Quality-factor and nonlinear properties of optical whispering-gallery modes. Physics Letters A. 1989;**137**(7):393-397. DOI: 10.1016/0375-9601(89)90912-2

[7] Sumetsky M. Mode localization and the Q-factor of a cylindrical microresonator. Optics Letters. 2010; **35**(14):2385-2387. DOI: 10.1364/OL.35.002385

[8] Knight JC, Cheung G, Jacques F, Birks TA. Phase-matched excitation of whispering-gallery-mode resonances by a fiber taper. Optics Letters. 1997; **22**(15):1129-1131. DOI: 10.1364/OL.22.001129

[9] Gorodetsky ML, Savchenkov AA, Ilchenko VS. Ultimate Q of optical microsphere resonators. Optics Letters. 1996;**21**(7):453-455. DOI: 10.1364/OL.21.000453

[10] Armani DK, Kippenberg TJ, Spillane SM, Vahala KJ. Ultra-high-Q toroid microcavity on a chip. Nature. 2003; **421**:925-928. DOI: 10.1038/nature01371

[11] Roselló-Mechó X, Delgado-Pinar M, Cruz JL, Díez A, Andrés MV. Measurement of UV-induced absorption and scattering losses in photosensitive fibers. Optics Letters. 2018;**43**(12): 2897-2900. DOI: 10.1364/OL.43.002897

[12] Armani AM, Kulkarni RP, Fraser SE, Flagan RC, Vahala KJ. Label-free, single-molecule detection with optical microcavities. Science. 2007;**317**(5839): 783-787. DOI: 10.1126/science.1145002

[13] Birks TA, Knight JC, Dimmick TE. High-resolution measurement of the fiber diameter variations using whispering gallery modes and no optical alignment. IEEE Photonics Technology Letters. 2000;**12**(2):182-183. DOI: 10.1109/68.823510

[14] Rivera-Pérez E. Microcavidades ópticas cilíndricas y sus aplicaciones [thesis]. Universidad Autónoma de San Luis de Potosí-IICO; 2015. pp. 1-84

[15] Schiller S, Byer RL. High-resolution spectroscopy of whispering gallery modes in large dielectric spheres. Optics Letters. 1991;**16**(15):1138-1140. DOI: 10.1364/OL.16.001138

[16] Arques L, Carrascosa A, Zamora V, Díez A, Cruz JL, Andrés MV. Excitation and interrogation of whispering-gallery modes in optical microresonators using a single fused-tapered fiber tip. Optics Letters. 2011;**36**(17):3452-3454. DOI: 10.1364/OL.36.003452

[17] Kenny RP, Birks TA, Oakley KP. Control of optical fibre taper shape. Electronics Letters. 1991;**27**(18): 1654-1656. DOI: 10.1049/el:19911034

[18] Sumetsky M, Dulashko Y. Radius variation of optical fibers with angstrom

accuracy. Optics Letters. 2010;**35**(23): 4006-4008. DOI: 10.1364/OL.35.004006

[19] Guan G, Arnold S, Otugen V. Temperature measurements using a microoptical sensor based on whispering gallery modes. AIAA Journal. 2006;**44**(10):2385-2389. DOI: 10.2514/1.20910

[20] Ma Q, Rossmann T, Guo Z. Temperature sensitivity of silica micro-resonators. Journal of Physics D: Applied Physics. 2008;**41**(24):245111. DOI: 10.1088/0022-3727/41/24/245111

[21] Rivera-Pérez E, Villegas IL, D´ ıez A, Andrés MV, Cruz JL, Rodríguez-Cobos A. Measurement of pump-induced temperature increase in doped fibers using whispering-gallery modes. IEEE Photonics Technology Letters. 2013;**25**(24):2498-2500. DOI: 10.1109/LPT.2013.2288865

[22] Wang Y, Xu CQ, Hong P. Thermal effects in kilowatt fiber lasers. IEEE Photonics Technology Letters. 2004; **16**(1):63-65. DOI: 10.1109/LPT.2003.818913

[23] Hadeler O, Ibsen M, Zervas MN. Distributed-feedback fiber laser sensor for simultaneous strain and temperature measurements operating in the radio-frequency domain. Applied Optics. 2001;**40**(19):3169-3175. DOI: 10.1364/AO.40.003169

[24] MJF D. Rare-earth-doped fiber lasers and amplifiers, revised and expanded. In: Optical Science and Engineering. Boca Raton: Taylor & Francis; 2001

[25] Davis MK, Digonnet MJF, Pantell RH. Thermal effects in doped fibers. Journal of Light-Wave Technology. 1998;**16**(6):1013-1023. DOI: 10.1109/50.681458

[26] Delgado-Pinar M, Villegas IL, Díez A, Cruz JL, Andrés MV. Measurement of temperature profile induced by the optical signal in fiber Bragg gratings using whispering-gallery modes. Optics Letters. 2014;**39**(21):6277-6280. DOI: 10.1364/OL.39.006277

[27] Poumellec B, Guénot P, Riant I, Sansonetti P, Niay P, Bernage P, et al. UV induced densification during Bragg grating inscription in Ge: SiO2 preforms. Optical Materials. 1995;**4**(4):441-449. DOI: 10.1016/0925-3467(94)00114-6

[28] Canning J, Åslund M, Hu PF. Ultraviolet-induced absorption losses in hydrogen-loaded optical fibers and in presensitized optical fibers. Optics Letters. 2000;**25**(22):1621-1623. DOI: 10.1364/OL.25.001621

[29] Arigiris A, Konstantaki M, Ikiades A, Chronis D, Florias P, Kallimani K, et al. Fabrication of high-reflectivity superimposed multiple-fiber Bragg gratings with unequal wave-length spacing. Optics Letters. 2002;**27**(15): 1306-1308. DOI: 10.1364/OL.27.001306

[30] Gagné M, Loranger S, Lapointe J, Kashyap R. Fabrication of high quality, ultra-long fiber Bragg gratings: Up to 2 million periods in phase. Optics Express. 2014;**22**(1):387-398. DOI: 10.1364/OE.22.000387

[31] Lauzon J, Thibault S, Martin J, Ouellette F. Implementation and characterization of fiber Bragg gratings linearly chirped by a temperature gradient. Optics Letters. 1994;**19**(23): 2027-2029. DOI: 10.1364/OL.19.002027

[32] Bertholds A, Dandliker R. Determination of the individual strain-optic coefficients in single-mode optical fibres. Journal of Lightwave Technology. 1988;**6**(1):17-20. DOI: 10.1109/50.3956

[33] Dragic PD, Ballato J, Morris S, Hawkins T. Pockels' coefficients of alumina in aluminosilicate optical fiber. Journal of the Optical Society of

America B: Optical Physics. 2013;**30**(2): 244-250. DOI: 10.1364/JOSAB.30.000244

[34] Bertholds A, Dandliker R. Deformation of single-mode optical fibers under static longitudinal stress. Journal of Lightwave Technology. 1987; **5**(7):895-900. DOI: 10.1109/JLT.1987.1075583

[35] Skelton SE, Sergides M, Patel R, Karczewska E, Maragó OM, Jones PH. Evanescent wave optical trapping and transport of micro- and nanoparticles on tapered optical fibers. Journal of Quantitative Spectroscopy and Radiation Transfer. 2012;**113**(18): 2512-2520. DOI: 10.1016/j.jqsrt.2012.06.005

[36] Lægsgaard J. Theory of surface second-harmonic generation in silica nanowires. Journal of the Optical Society of America B: Optical Physics. 2010;**27**(7):1317-1324. DOI: 10.1364/JOSAB.27.001317

[37] Roselló-Mechó X, Delgado-Pinar M, Díez A, Andrés MV. Measurement of Pockels' coefficients and demonstration of the anisotropy of the elasto-optic effect in optical fibers under axial strain. Optics Letters. 2016;**41**(13):2934-2937. DOI: 10.1364/OL.41.002934

[38] Jülich F, Aulbach L, Wilfert A, Kratzer P, Kuttler R, Roths J. Gauge factors of fibre Bragg grating strain sensors in different types of optical fibres. Measurement Science and Technology. 2013;**24**(9):094007. DOI: 10.1088/0957-0233/24/9/094007

[39] Pöllinger M, O'Shea D, Warken F, Rauschenbeutel A. Ultrahigh-Q tunable whispering-gallery-mode microresonator. Physical Review Letters. 2009;**103**:053901. DOI: 10.1103/PhysRevLett.103.053901

[40] Linslal CL, Kailasnath M, Mathew S, Nideep TK, Radhakrishnan P, Nampoori VPN, et al. Tuning whispering gallery lasing modes from polymer fibers under tensile strain. Optics Letters. 2016;**41**(3):551-554. DOI: 10.1364/OL.41.000551

[41] Lide DR. CRC Handbook of Chemistry and Physics. 84th ed. Boca Raton: CRC Press; 2003

[42] Shackelford JF, Alexander W. CRC Materials Science and Engineering Handbook. 3rd ed. Boca Raton: CRC Press; 2001